Smart Geospatial Practices and Applications in Local Government

Smart Geospatial Practices and Applications in Local Government

An Altogether Different Language

David A. Holdstock

CRC Press
Taylor & Francis Group
Boca Raton London New York

CRC Press is an imprint of the
Taylor & Francis Group, an **informa** business

CRC Press
Taylor & Francis Group
6000 Broken Sound Parkway NW, Suite 300
Boca Raton, FL 33487-2742

First issued in paperback 2022

© 2020 by Taylor & Francis Group, LLC
CRC Press is an imprint of Taylor & Francis Group, an Informa business

No claim to original U.S. Government works

ISBN 13: 978-1-03-247497-7 (pbk)
ISBN 13: 978-1-138-05457-8 (hbk)
ISBN 13: 978-1-315-16664-3 (ebk)

DOI: 10.1201/9781315166643

Visit the Taylor & Francis Web site at
http://www.taylorandfrancis.com

and the CRC Press Web site at
http://www.crcpress.com

To my wife:

On my desk stands a framed quote. It reads "You are never too old to set another goal or to dream a new dream" C. S. Lewis

To Natalie, Amelia, Tosh, and Bea:

"One cannot divine nor forecast the conditions that will make happiness; one only stumbles upon them by chance, in a lucky hour, at the world's end somewhere, and holds fast to the days, as to fortune or fame."

Willa Cather, "Le Lavandou," 1902

I love you

Mr. David Andrew Holdstock

Pseudonym, Dr. Woodstock

Contents

Acknowledgments

I am infinitely pleased and proud to be associated with such a wonderful group of creative and energetic people at Geographic Technologies Group (GTG). Thanks go to my business partner Mr. Curt Hinton, and Mr. James Kelt, Mr. Jason Marshall, Mr. Jonathan Welker, Mr. Rives Deuterman, Ms. Dawn Reim, for all her logistical talent, and all our GIS Managers and Specialists. Thanks to Mr. Cameron Higgins, editor in chief, who came through like a thoroughbred racehorse! Special thanks to Ms. Jessica Parker for her organizational ability and management style, and to the gifted Ms. Emely Srimoukda who has the power and artistic talent to bring a book to life.

Here's to all those people in local government who have inspired me and didn't even know it. Thank you for being an important part of my story.

- City of Berkeley, California – Savita Chaudhary and Cristi Delgado
- City of Hobart, Indiana – Bob Fulton and Tim Kingsland
- City of Irvine, California – Rebecca Bridgeford and Mike Sherran,
- City of Mississauga, Ontario, Canada – John Imperiale
- City of Nanaimo, British Columbia, Canada – Mark Willoughby
- City of Roswell, Georgia – Patrick Baber
- City of West Hollywood, California – Eugene Tsipis
- City of Wilson, North Carolina – Will Aycock
- Town of Windsor, California – Carl Euphrat
- City of Vancouver, Washington – Eugene A. Durshpek
- Volusia County, Florida – Al Hill
- Yukon Energy, Yukon, Canada – Shannon Mallory
- City of Markham, Ontario, Canada – Ewan Burke
- City of Morgan Hill, California – Pam Van der Leeden
- County of Spotsylvania, Virginia – Jane Reeve and Rich Maidenbaum
- Contra Costa Water District (CCWD), California – Richard Broad
- Lane Council of Government (LCOG), Oregon – David Richey
- Gwinnett County, Georgia – Sharon Stevenson

They say show me your friends, and I'll show you your future. I offer an enormous thank you to my dear lifelong friends, for without them this book would have been completed two years ago. To Mr. Scott Mahoney – a remarkable, successful and extraordinary honest man, with a sharp wit and a low tolerance for time-wasting. Mr. Paul Holdstock – indelible and rather handsome for the funniest man alive; Mr. David Paulson, Esq. – poised with the intellectual capacity of a giant; Mr. Dan Higgins – a constant workhorse of a man who has always been there for me; Mr. Mike Doerfler – Zeus of a man with talent, strength, and humor, and some say a tortured soul with a glance as trenchant as a heavy axe; Mr. David Cozart – charm, wit, and the most tranquil man you could ever meet; Mr. Bob Hofstadter – who has the good fortune of being from the south; Mr. Richard Wilson – a successful sportsman and business man with a kind soul and a penchant for fun and laughter; Mr. Gary Hartley – an athlete, a gentleman, with an unusual fear of flying; Mr. Keith Willis – an honest, kind, and sincere friend with a firm sense of his own mind and a rather thrilling way with words; Mr. John Risman – a true Jane Austin hero, who's kindness and true loyalty are rare and, once secured, are secured forever; Mr. Steve Thomas – a man of spirit and amiability, no one may surpass him; Dr. Stephen Mcculley – one of the most trustworthy gentlemen in our circle, added to his respectable qualities is a very sensible taste in clothes; and Richard "Buster" Musto – in kindness, judgment, and appearance, everything that a gentleman ought to be. And thanks goes to a fella that had an enormous influence on me throughout my life, my older brother Mr. Stephen Holdstock.

I would also like to thank my favorite people in the world Pete and Pickel Tannenbaum and Charles and Jenny Winston. There are some people you just can't be without.

And to my wife, who is bound to my soul with hoops of steel. "I cannot fix the hour, or the spot, or the look or the words, which laid the foundation." Jane Austen.

Preface

Cities have the capability of providing something for everybody, only because, and only when, they are created by everybody.

——Jane Jacobs
The Death and Life of Great American Cities

The chances of you getting mugged in your home town are far greater than you getting mugged in New York City. How can that be? Well, because you actually live in your home town. Sometimes the obvious escapes us all.

When Arthur Jennings heard that a new United States Geological Survey (USGS) placed his home in Iowa rather than Wisconsin, Mr. Jennings expressed his delight at this as he wouldn't have to suffer the long Wisconsin winters anymore.

Two friends set out on a hike across Western Australia. On the afternoon of the first day, one fella looks out across the open landscape and says "Hey mate, look at those beautiful fields of barley." His traveling companion replies, "No mate, that's wheat." Early, the next morning the wheat-seeing hiker catches the barley-seeing hiker heading out of camp on his own. His friend asks, "Where are you going?" His companion replies "I'm leaving, too much bloody arguing" (Clive James, CBE).

Working with so many talented government professionals fuels my enthusiasm about the future of geospatial technology in our towns, cities and counties as it relates to high performance, business realization, culture and the dramatic opportunities presented by software applications.

We must understand geography and science. We must have open and honest dialogue about our future urban centers. We do not have time to argue. Our challenge is to understand the growing geospatial ecosystem of software and solutions and embrace a new mindset for the future. Indeed, we must learn "An Altogether Different Language" that includes:

- Geo-spatial Technology
- Geosmart Government
- A Connected Technology Ecosystem
- A System of Systems
- High Performance Organizations (HPO)

- Business Realization Planning (BRP)
- Organic and and Continuous Management and Improvement
- Key Performance Indicators (KPI)
- Sustainability and Resilience Indicators
- Environmental, Economic and Social Factors
- GIS Software Solutions

We must also focus on looking forward. As Sir Fred Hoyle said in his after dinner speech at a lunar science conference in Houston in 1970, "Everything that is past died yesterday, says the poet. Everything in the future was born today."

About the Author

Mr. David A. Holdstock, BA, MS, a geographic information systems (GIS) Professional and chief executive officer, co-established and incorporated Geographic Technologies Group, Inc. in 1997. Mr. Holdstock is a GIS practitioner and a leading expert in developing enterprise and sustainable GIS strategic implementation plans for towns, cities and counties. Over the past twenty-five years, Mr. Holdstock has planned, designed and coordinated the adoption and implementation of GIS technology for over 200 local government organizations. He has published many articles on GIS strategic planning for local government. He has conducted hundreds of workshops, seminars and discussions on GIS implementation. His previous work experience has included being a GIS manager for the world's leading transportation engineering company in New York and a GIS director for a research institute at North Carolina State University.

1

Introduction

The true sign of intelligence is not knowledge but imagination.

– Albert Einstein

Introduction

Conventional wisdom says that wide-ranging, multi-disciplinary events and a myriad of new technological pathways influence our current geographic information system (GIS) environment. In this introduction, I offer a brief history of GIS technology: how GIS professionals influenced the industry and how GIS technology reinvented local government. I discuss how rapid technological evolution rather than technological paradigm shifts influenced the geospatial industry. I also introduce and explain the hot topic of smart, sustainable, resilient and high-performing local government organizations and briefly discuss the significance of business realization planning. Lastly, I volunteer descriptions of our complex geospatial future and the myriad ways a new GIS maturity model would benefit us all.

Before we dive headlong into the world of GIS and geo-smart government, I just want to say that, for the record, the science of GIS technology is fundamentally about connecting people with other people and data in service of informed decision making. Our evolving geospatial ecosystem connects and engages people, transforms communities and modernizes our world. An interconnected world will drive innovation, promote our economy, enhance education and promote social change.

A Brief GIS History: The GIS Professional as an Agent of Change

The progression of GIS and geospatial technology stays in step with other related technology and management styles. The technology pioneers of the 1960s created the blueprint of modern GIS. Next, the GIS Managers of the

1980s and 1990s directed the technology's evolution. Finally, from the turn of the millennium to the present day, the demands of GIS coordinators have reshaped the technological field. In the near future – between 2020 and 2040 – our future geographic information officers (GIO) will define and be defined by a very different smart geospatial environment. We must understand our history if we are to successfully prepare for our future. I believe that **the future is always the present. We react, we do not predict**.

In the summer of 2018, I was invited to be the opening keynote speaker at the Elevations Geospatial Summit in Wyoming. The title of my presentation was "GIS: Yesterday, Today, and Tomorrow" (Holdstock, 2018). I presented my paper with the following disclaimer: my twenty-five years of local government GIS strategic planning experience and twenty-six years of British deadpan humor shaped and fashioned everything I had to say. What followed were personal, albeit first-hand empirical observations regarding the geospatial industry since 1993. The outline of my speech was simple:

- A Brief GIS History: The GIS Professional as an Agent of Change
- 1960 to 1980: Proof of Concept: GIS Pioneers
- 1980 to 2000: Desktop, Analysis, and Projects: GIS Managers
- 2000 to 2020: Enterprise, Strategic, and Scalable: GIS Coordinators
- 2020 to 2040: Smart, Resilient and Sustainable: Geographic Information Officers (GIO)

Figure 1.1 illustrates a timeline of GIS and Geospatial technology compartmentalized into distinct periods of GIS development.

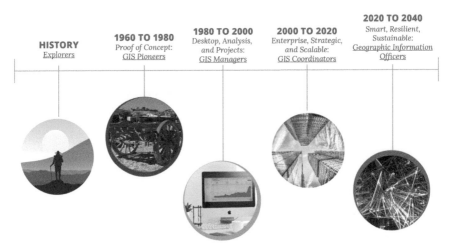

FIGURE 1.1
Timeline of GIS and geospatial technology.

I began my keynote speech with the following observation: what I love dearly about local government GIS professionals is that so many of them utterly and completely believe that life is changeable, definable and improvable through the use of geospatial technology. That their work is something that robots would never be able to do! In my attempt to do a bang-up professional job with my keynote speech, I ended up reading a lot about the history of related geospatial technology that may, in some small way, have influenced our geospatial world today. My preparation included researching technological successes and technological failures over the last 100 years. I was amused by the incredible inaccuracy of technology predictions about the future. After all, one of my goals in this book is to predict the future of geospatial technology after 2020. It must be part of the human condition to routinely prognosticate, since it seems we are constantly trying to predict the future of everything. As we all know, it is relatively easy to predict tomorrow's weather, but far more difficult to anticipate events farther into the future. If we can accurately foretell the future of our geospatial technology in local government, we will be able to make solid decisions and successfully prepare to meet our vision, goals and objectives.

As Sir Isaac Newton said in 1675, "if I have seen further it is by standing on the shoulders of giants." As geospatial industry professionals, we should thank all the giants that have changed the course of geospatial history beginning with the men and women of old who furthered **exploration, navigation, cartography, ship building and development of the printing press, compass, lightbulb, Morse code, radar, telegraph and camera.** Then, of course, you have notables like **Copernicus, Galileo, Kepler, Einstein** who advanced human knowledge as a whole. We can't forget the **Viking civilization** for spearheading a form of immigration (invasion) and the **Romans** for establishing a template for civilization that includes straight roads, drainage, sewers, viaducts and baths. Note that the Romans gave us the ideas, concepts, and formulas and innovative urban grid layouts still used to control vast amounts of land, housing and buildings.

Fast forward to a few thousand years to the well-documented beginnings of geospatial analysis by **Charles Picquet** in 1832 and **Dr. John Snow** in 1854. Charles Picquet created a heat map representation of the cholera epidemic in Paris by representing the forty-eight districts with different halftone color gradients. Led by Dr. John Snow, a similar mapping analysis depicted cholera deaths in London using points on a map. Although Dr. Snow came twenty-two years after Picquet, he still managed to have a London pub named after him. Those British!

Considering that this book is, in part, about my predictions for the future of geospatial technology, I would be remiss not to mention two of the most famous prognosticators in the history of mankind. In 1490, **Leonardo da Vinci** talked about and created drawings of planes, parachutes, helicopters, hang gliders and the adding machine. In 1883, **Jules Verne** predicted glass skyscrapers, air conditioning, television, elevators and fax machines.

But with every accurate technological prediction made between 1800 and 1960, came a boatload load of profoundly wrong technological forecasts. Here are some of the remarkable highlights:

- *1876: "This 'telephone' has too many shortcomings to be seriously considered as a means of communication."*
- *1899: "Everything that can be invented has been invented."*
- *1903: "The horse is here to stay but the automobile is only a novelty – a fad."*
- *1921: "The wireless music box has no imaginable commercial value. Who would pay for a message sent to no one in particular?"*
- *1926: "While theoretically and technically television may be feasible, commercially and financially it is an impossibility."*
- *1932: "There is not the slightest indication that nuclear energy will ever be obtainable. It would mean that the atom would have to be shattered at will."*
- *1936: "A rocket will never be able to leave the Earth's atmosphere."*
- *1949: "Where a calculator on the ENIAC is equipped with 18,000 vacuum tubes and weighs 30 tons, computers of the future may have only 1,000 vacuum tubes and perhaps weigh one and a half tons."*
- *1957: "I have traveled the length and breadth of this country and talked with the best people, and I can assure you that data processing is a fad that won't last out the year."*

In this eighty year period, humanity went from the age of electricity in the 1800s to the age of radio from 1900 to 1930 to the age of instruments between 1930 and 1945 and ultimately, the start of the computer age in 1945. A lot of heavy lifting went on before 1960! All of these developments were precursors to the GIS systems we know today.

1960 to 1980: Proof of Concept: GIS Pioneers

In 1961, President John F. Kennedy made a promise to the world: by the end of the decade, the United States of America would put a man on the moon and bring him safely home. On, July 20, 1969, Apollo 11 astronaut Neil Armstrong became the first person to step onto the surface of the moon, six hours after the lunar landing. Twenty minutes later, just after midnight on July 21, Buzz Aldrin joined Armstrong on the lunar surface. The pair spent approximately two and a quarter hours together outside the spacecraft, collecting the 47.5 pounds of lunar material they would eventually bring back to Earth. NASA didn't achieve the "giant leap for mankind" without Federal agencies' efforts to create detailed maps and charts of the lunar surface. Cartographers compiled extensive maps by using computers to piece together disparate images of the moon landscape. If you still don't believe that all our space explorations will require GIS and geospatial solutions just watch Kate Mara and Matt Damon in the 2015 movie *The Martian* (Scott et al., 2015).

Back on earth, the twenty year period between 1960 and 1980 marked significant events in the evolution of GIS. 1960–1975 was the age of computers. Recall that (incredibly) the computer NASA used by the Apollo 11 program was by today's standards no more powerful than a pocket calculator. Two people stand out in the early 1960s GIS landscape: Mr. Ian McHarg and Dr. Roger Tomlinson, who later become known as the "Father of GIS." The year 1962 saw the development of the world's first true **Operational GIS** in the city of Ottawa, Ontario, Canada. Developed by Dr. Roger Tomlinson and established by the Federal Department of Forestry and Rural Development, Ottawa's revolutionary GIS system was called the Canada Geographic Information System (CGIS). The CGIS stored, analyzed and manipulated data collected for the Canada Land Inventory (CLI). The CLI was an initiative that determined land capabilities for rural Canada by mapping information about soils, agriculture, recreation, wildlife, waterfowl, forestry and land use. In her article "History of GIS," Caitlin Dempsey, editor of GIS Lounge, explains that "by the 1960s the nuclear arms program had given rise to hardware and mapping applications and the first operational GIS had been launched in Ottawa" (Dempsey, 2018).

We must also mention Mr. Jack Dangermond and his wife. It's about time that someone – as my mother always said – gives them credit where credit is due! Mr. Dangermond needs to take his rightful place in the history of GIS and geospatial science. An enormous number of people, organizations, businesses and governments have clambered up onto the shoulders of this giant of a man and pushed the field further. Remember, behind every great man there is a great woman. Jack Dangermond, born 1945, is an American *environmental scientist and businessman* who co-founded the *Environmental Systems Research Institute* (Esri) with his wife Laura in 1969. If Roger Tomlinson is the "Father of GIS," then Jack and Laura Dangermond are the wunderkinds that follow. They are the greatest innovators of geospatial science in the history of humankind and will be long outlasted by their accomplishments.

This early period of GIS – or the "proof of concept" era – was characterized by significant increases in environmental awareness coupled with a technologically enhanced environmental movement that deployed new tools such as NASA images, light-emitting diode (LED) lights, the personal computer (PC) in 1977, television, satellites, drones, **radio-frequency identification tags** (RFID), hand-held calculators by Texas Instruments, Apple and graphical user interface (GUI). Due, in part to the environmental movements' success, new environmental laws were passed in 1976. Furthermore, in the 1970s, the Harvard Laboratory Computer Graphics was recognized as the first Vector GIS (*ODYSSEY*). Computer technology progressed tremendously during this period:

- Main frame to mini computers
- Mini computers to PC
- PC to mobile devices
- Batch processing to interactive
- Command line to GUI

Many GIS professionals say that the development of digital mapping software was born in the 1970s from the need to convert hardcopy maps into a digital format. The early federal systems ran on large mainframe computers. The Census Bureau and the United States Geological Survey (USGS) are prime examples of 1970 geospatial initiatives. In spite of some tremendous advances in technology between 1960 and 1980, the period still generated remarkably incorrect predictions about the future:

- *1960: "There is practically no chance communications space satellites will be used to provide better telephone, telegraph, and television or radio service inside the United States."*
- *1966: "Remote shopping, while entirely feasible, will flop."*
- *1968: "There would be a major food shortage in the US. In the 1970s ... hundreds of millions are going to starve to death."*
- *1974: "It will be years – not in my time – before a woman will become Prime Minister." – Margaret Thatcher*
- *1970: "By 1985, air pollution will have reduced the amount of sunlight reaching earth by one half."*
- *1970: "Air pollution ... is certainly going to take hundreds of thousands of lives in the next few years alone."*
- *1970: "Scientists have solid experimental and theoretical evidence to support ... the following predictions: In a decade, urban dwellers will have to wear gas masks to survive air pollution."*

1980 to 2000: Desktop, Analysis and Projects: GIS Managers

During the "age of electronics" that lasted from 1975 to 1995, we were first introduced to the *World Wide Web (WWW)* and everything that entailed: the *internet, virtual reality, smart phones, social media, optical sensors, Hubble Space Craft* in 2000, *texting,* and *global positioning system* (GPS) technology. These advances were influenced by President Ronald Reagan in 1983 and again in 2000 by President Bill Clinton. This period in GIS history includes the "Backroom GIS Specialists." Rumor said that one had to push pizza under their office door and hold on tight in hopes that these wizards could make the GIS technology work. Of course, GIS in local government was characterized at this time by a decentralized GIS governance model. That is to say: individuals and departments were "going it alone" when they worked with and discovered the benefits of GIS. This era also brought forth desktop GIS, rudimentary spatial analysis and siloed databases. In 1982 we welcomed in Esri's ARC/INFO for mini computers, and then in 1986 PC ARC/INFO was launched. It was a busy time for all GIS professionals. The following is list of release dates for Esri software:

- 1982 – Arc INFO coverages
- 1986 – PC Arc INFO coverages
- 1991 – Arc INFO grid
- 1991 – Arc CAD
- 1991 – Arc View shapefiles
- 1994 – Arc Storm
- 1995 – Arc Spatial Database Engine (SDE) – Relational Database Management System (RDMS)
- 1996 – Map Objects
- 1997 – Arc View Internet Map Server (IMS)
- 1999 - Arc IMS – Map Services
- 1999 – Models and Extensions

We must also note that open-source (**open-source** software is **source code** that can be freely accessed and modified) GIS can be traced back to the U.S. Department of the Interior in the late 1970s. Along with software advances, something revolutionary happened during the 1980s: the GIS Analyst and the GIS Managers were born. This event marks a milestone in the history of GIS governance.

Of course, no era would be complete if humanity did not make some astronomically wrong predictions for the future. Between 1980 and the year 2000 it was predicted that:

- *1980: "There is no reason for any individual to have a computer in his home."*
- *1981: "No one will need more than 637 KB of memory for a personal computer. 640 KB ought to be enough for anybody."*
- *1981: "Cellular phones will absolutely not replace local wire systems."*
- *1989: "We will never make a 32-bit operating system."*

- *1992: "The idea of a personal communicator in every pocket is a pipe dream driven by greed."*
- *1995: "I predict the Internet will soon go spectacularly supernova and in 1996 catastrophically collapse."*

2000 to 2020: Enterprise, Strategic and Scalable: GIS Coordinators

The turn of the millennium ushered in the true "age of the internet." In this era, GIS has been defined by **common operating pictures (COP), improved RFID tags, fiber optics, crowdsourcing,** renewed looks at **open source, mobile GIS** and **hybrid GIS governance model** in local government (a management model that combined the best of decentralized model and the centralized model), **web solutions, dashboards, enterprise and scalable geodatabases, ArcGIS Server** and the fading line **between cell phone and PCs.**

In January 2004, the Urban and Regional Information Systems Association (URISA) created the GIS Certification Institute (GISCI), which introduced GIS professional (GISP) certification as a way to promote competent and ethical professional practices. Since then, nearly five thousand practitioners from across the United States and around the world have earned GISP certification. Yet, as always, we still find technological naysayers:

- *2003: "The subscription model of buying music is bankrupt. I think you could make available the Second Coming in a subscription model, and it might not be successful."*
- *2007: "There's no chance that the iPhone is going to get any significant market share."*

During this period we witnessed the birth of the early GIS coordinator. This position was characterized by what I call the "multiple-personality GIS professional." These coordinators functioned as GIS controllers, comedians, thinkers and artists. We essentially gave birth to many types of GIS coordinators. People entering the GIS world came from a variety of backgrounds, including geography, planning, landscape architecture, natural resource management, engineering, psychology and many other disciplines.

To offer a full picture of the various GIS applications that defined the last nineteen years, I reviewed the last forty GIS Strategic Plans created for towns, cities and counties by Geographic Technologies Group – a world leader in geospatial planning. I also used the 1000 GIS Applications & Uses (2019) from the website, GISGeography.com. My goal was to identify the myriad of GIS uses in local government. Though it seemed that we had entered a period of "data rich and application rich organizations," I knew that wasn't quite true. Figures 1.2 through 1.9 list GIS software uses and applications in local government.

2020 to 2040: Smart, Resilient and Sustainable: Geographic Information Officers (GIO)

The years 2020 to 2040 will usher in the geospatial city of the future or the "smart city." As we discuss this period, consider the words of Harvard

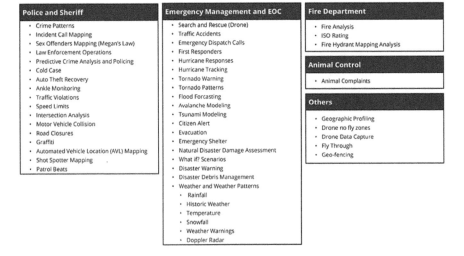

FIGURE 1.2
Local government GIS software uses and applications – Public safety and law enforcement.

Local Government GIS Software Uses and Applications
• Local Government
• Telecommunications
• Broadband Service Provider
• Fiber Optic Cable
• Cell Towers
• OSS/BSS Data
• Cellular Coverage Analysis

FIGURE 1.3
Local government GIS software uses and applications – Telecommunications.

FIGURE 1.4
Local government GIS software uses and applications – Resources, forest preserves, open space and parks and recreation.

economist John Kenneth Galbraith when he said, "There are two kinds of forecasters: those who don't know and those who don't know that they don't know." I count myself among the latter. However, we must also remember Helen Keller's famous observation that, "no pessimist ever discovered the secrets of the stars or sailed to an unchartered land or opened a new heaven to the human spirit."

With the previous words firmly in mind, I assert that the **smart city** will characterize the GIS landscape between 2020 and 2040. Figure 1.10 illustrates the smart city technology that will become a common place in the near future.

I believe in the following words of Emeritus Professor of Geography at the University of California, Santa Barbara, Dr. Michael Goodchild: "GIS and geospatial technology will eventually move outside the constraints of the map to understand relationships among human behavior across space and time."

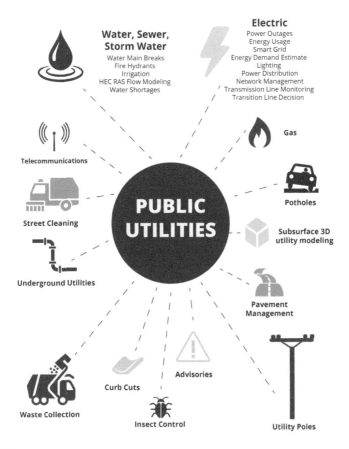

FIGURE 1.5
Local government GIS software uses and applications – Public utilities.

I predict that the next twenty years will see the birth of the geographic information officer (GIO) and with it a new level of geospatial services and support.

Figure 1.11 shows a new philosophy that includes four distinct shifts in the way we think.

The role of the GIO will include the following listed in Figure 1.12.

What will happen is the growth of a total geospatial ecosystem or what I call "Bespoke Design Solutions" (BDS). Customizing tailored geospatial solutions for local governments will be a primary focus for many a GIS professionals. The BDS will focus on integrating the Esri ecosystem of software solutions into specific business applications. Browsers will be combined with mobile tools, dashboards and business widgets to tailor unique solutions for local government.

I recently returned from the Esri Business Partner conference. They are focused on all of the technological initiatives but have not honed in on

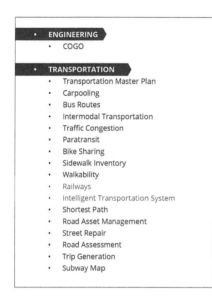

FIGURE 1.6
Local government GIS software uses and applications – Public works.

culture, value and performance. The future is not just technology. Rather, it is the way technology intersects with the next generation's philosophy and mindset.

Today's Local Government Departments and Functions

To understand the changing structure of local government today, we must pay special attention to geospatial functions and applications. Geospatial technology influences local governmental operations now more than ever. Local government's use of geospatial technology is vastly different than it was a mere decade ago. It goes without saying that technology will look wildly different in twenty years.

According to some simple Google searches, there are over 3,000 counties and county-equivalents in the United States (Note Louisiana and Alaska have county equivalent subdivisions called parishes and boroughs). The specific governmental powers of counties vary widely between the states. Most counties include unincorporated areas with variously designated cities, towns, villages, boroughs and municipalities. Within the United States, there are approximately 89,000 local governments including 19,522 municipalities, 16,364 townships, 37,203 special districts and 12,884 independent school districts. The combined municipality and township would equal about 35,000

PLANNING AND ZONING DEPARTMENTS

- Archeological Sites
- View shed Analysis and Line of Sight
- Cultural Heritage
- 3D Archeological
- Noise
- Crowd Simulation
- Solar Exposure
- Urban Modeling City Engine (Esri)
- Pedestrian Modeling
- Parking
- Space Utilization
- Addressing
- Commercial Space Availability
- Land for Development
- Land Use Change
- Homeless Shelters
- Cemetery Mapping
- Construction
- Tourism Planning
- Historic Street View
- Aviation flight path
- Multi-modal mapping
- Easements

ECONOMIC DEVELOPMENT

- Economic Viability
- Market Analysis
- Foreclosures
- Retail Site Selection
- Real Estate Analysis
- Socioeconomic data

BUILDING AND INSPECTIONS

- Building Permits
- Historic Buildings
- Building Constraints
- Future Development Plans
- Dam Site Inspections

CODE ENFORCEMENT

TAX ASESSORS

- Tax collection
- Tax Parcel Viewer
- Parcel Fabric

INFORMATION TECHNOLOGY DEPARTMENT

- Storytelling
- Social Media Mapping
- Demographic research
- Urban traffic air pollution
- Open Data

PUBLIC INFORMATION OFFICER

PARTICIPATOR GIS

WELLS

SHIPPING ROUTES

PIPELINE INFRASTRUCTURE

SINKHOLES

FLOODPLAIN

LANDFILL SITE SELECTION

FIGURE 1.7

Local government GIS software uses and applications – Land and Information Management.

FIGURE 1.8
Local government GIS software uses and applications – Public administration.

FIGURE 1.9
Local government GIS software uses and applications – Public service.

cities. It is estimated that about half of the U.S. population resides in a city or town with fewer than 25,000 people. We should confirm all of these estimates with the 2020 US Census.

I have been roaming around North America for twenty-five years listening carefully to the creative ideas and thoughts of local government professionals. The greatest compliment I can give to these people comes in the words of a great American, Theodore Roosevelt (1910), "the credit belongs to the person who is actually in the arena, whose face is marred by dust and sweat and blood; who strives valiantly; so that their place shall never be with those cold and timid souls who neither know victory nor defeat." I believe that local government practitioners are the people Teddy speaks of. They have shaped and reshaped the GIS industry by evolving geospatial business practices and pushing the geospatial envelope. It takes great imagination to plan, design and implement new ways of conducting business. We owe an incredible debt of gratitude to the men and women of local government, because they are the great incubators of GIS ideas. For this reason, coupled

Smart City Technologies				
A System of Systems	Smart Connected Assets and Systems	Connected Smart City Ecosystem	High Performance Organizations (HPO)	Resilient Communities
Sustainable Communities	Regionalizaton of GIS	Satellite Data	Fusion Centers	Smart Devices
Big Data Analytics	Applied Predictive	Analytics to produce new insights as well as new products, Cloud GIS	Citizens as sensors	Wearable GIS
Geographic Information Officers (GIO)	Sensors and Cameras	Interoperability	New IoT Platform	More Spatially literate society
Driverless Cars	Drone Technology	Internet Glasses and Contact Lenses	Micro Chips	Computer Wall Screens
Flexible Electronic Paper	Virtual Worlds			

FIGURE 1.10
Smart city technologies.

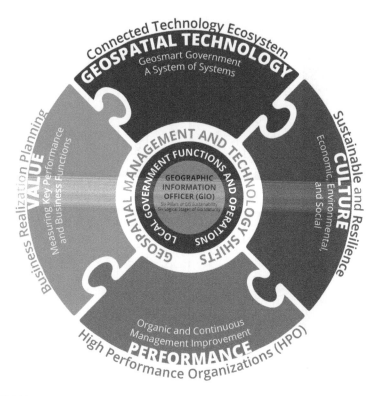

FIGURE 1.11
Smart geospatial practices and applications in local government: geospatial technology, culture, value and performance.

with the fact government constantly aims to spend their limited tax dollars wisely and effectively, our local governments remain in a constant state of technological improvement.

However – and there is always a however – we should also recognize the enormous gaps and opportunities regarding the use of GIS and geospatial technology that persist in local government to this day. These gaps result from towns', cities' and counties' enormous appetite for the innumerable uses of geospatial technology and the fact each government organization finds itself at a different maturity level regarding their geospatial evolution. On occasion, even large organizations have taken a few steps backwards in their geospatial journey. For example, the economic downturn in 2008 deeply impacted the fast-moving realm of geospatial technology.

According to author Romy Varghee of *Bloomberg Businessweek*, technology in local government is held together with "chicken wire and duct tape ... municipal offices across the country are struggling to do their jobs with obsolete gear." Police departments, emergency services, courts, transit departments and more are working with outdated software solutions. This situation requires immediate attention.

Geographic Information Officer (GIO) Job Descriptions				
Resilience Planning	Cyber Security	Augmented Reality and Visualization	Real Time Data and Monitoring	Indoor Maps
Crowdsourcing	Sustainable Planning	Regionalization	High Performance Organizations (HPO)	Total GIS Ecosystem
Customer (Residents) Connection	Smart City Initiative	Interoperability - System of Systems	Drone Technology	Privacy and Security
Advanced Predictive Analytics	Agnostic Technology Platform	Bespoke Design Solutions	3D Mapping ArcGIS HUB Administration	

FIGURE 1.12
Geographic information officer (GIO) job descriptions.

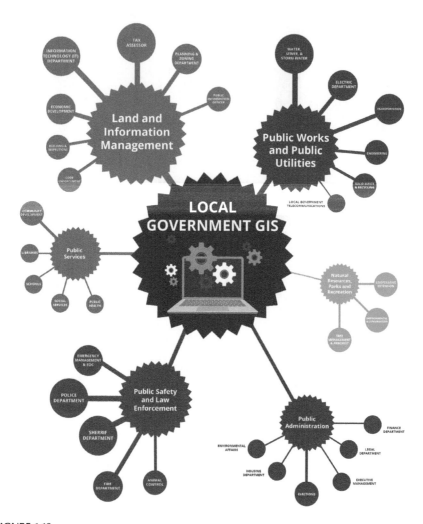

FIGURE 1.13
Local government departments organized into groups of geospatial users.

The evolution of GIS over the last thirty years has been complex, dynamic and particularly interesting on the community level. It seems that each and every department and division of local government use geospatial machinery, equipment, expertise and know-how to a lesser or greater extent. The following Figure 1.13 is a list of local government departments organized into groups of GIS or geospatial users.

Group 1: Heavy Hitters

- Tax Assessor
- Police Department

- Sheriff Department
- Water, Sewer and Storm water
- Information Technology (IT) Department
- Geographic Information Systems (GIS) Division

Group 2: Technical Specialists

- Planning and Zoning Department
- Electric Department
- Emergency Management and Emergency Operations Center (EOC)
- Fire Department
- Economic Development
- Transportation

Group 3: Easy Enthusiasts

- Animal Control
- Engineering
- Building and Inspections
- Code Enforcement
- Public Information Officer
- Parks and Recreation Departments
- Tree Management and Arborist
- Environmental and Conservation
- Cooperative Extension
- Elections
- Community Development
- Solid Waste and Recycling

Group 4: The Outliers

- Executive Management
- Legal Department
- Finance Department
- Housing Department
- Environmental Affairs
- Libraries
- Schools

- Social Services
- Public Health
- Telecommunications

Figure 1.14 graphically illustrates the use of GIS and geospatial technology in each department of local government. This graphic is a generalization and may differ between organizations.

This section is all about smart GIS departmental practices and applications. It covers seven key areas that make up the thirty-four departments within our local government, including:

1. **Public Safety and Law Enforcement:** Police, Sheriff, Emergency Management and EOC, Fire Department, Animal Control

2. **Public Works and Public Utilities:** Water, Sewer, Storm Water, Solid Waste and Recycling, Engineering, Transportation, Electric, Telecommunications

3. **Land and Information Management:** Planning and Zoning Department, Economic Development, Building and Inspections, Code Enforcement, Tax Assessor, Information Technology Department, Public Information Officer (PIO)

4. **Natural Resources, Parks and Recreation:** Tree Management and Arborist, Environmental and Conservation, Cooperative Extension

5. **Public Administration:** Executive Management, Legal Department, Finance Department, Housing Department, Environmental Affairs, Elections

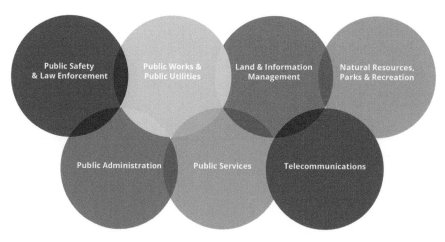

FIGURE 1.14
The Use of GIS technology in each department of local government.

6. **Public Services:** Library, Schools, Public Health, Social Services, Community Development

7. **Telecommunications:** Local Government Telecommunications, Broadband Service Providers

My experience tells me that, on the one hand, local government organizations excel at using GIS technology. On the other hand, though, these same local government organizations often have gaps to fill, challenges to address and barriers or pitfalls to overcome. That withstanding, I believe that given the myriad uses for geospatial tools in local government and rapid evolution of geospatial sciences, GIS technology will remain in a perpetual state of underutilization.

Twenty-three years ago in 1995, geographers and professors of spatial science, Stewart Fotheringham and Peter Rogerson stated, "One of the observations that is made repeatedly is that there is a mismatch between the spatial analytical capabilities of the research community and the applied tools available in GIS and in use by practitioners" (Fotheringham and Rogerson, 1995/2014). They went on to say that non-geographers using GIS do not think in spatial terms, that researchers found it hard to justify the applied use of spatial tools for real-world problems, and that many spatial analytics tools are limited by the size of the market.

How true is this today? Have we changed the way we think about space and GIS? Well, considering that the "where" factor has become standard integrated practice in phones, watches, motor vehicles and thousands of software applications one could fairly believe the answer is an emphatic "yes!" We have dramatically changed the way we think about space and information systems technology. Local governments have totally revamped the ways they consider and utilize geography.

GIS technology continues to transform government operations. Over the last five years, GIS has become dramatically more accessible, easier to implement and more enjoyable to use. There is overwhelming evidence that GIS technology improves efficiency, increases productivity and automates traditionally manual processes. Governments today aren't just seeking new answers, they are asking entirely different questions. These new queries relate to social equity, sustainability, resilience and democratization of GIS.

So the question is not whether GIS offers us a return on investment (ROI). That argument is done and dusted (note: the *Oxford English Dictionary* defines this phrase as meaning "completely finished or ready"). The real question is: how is geospatial technology changing local government operations?

Even though the evidence indicates that governments are realizing the benefits of GIS, we must concede the difficulty of measuring success in this realm. Thus, we must consider the application of business realization planning that introduces mechanisms to monitor, track and measure success.

Evolution and Maturity of GIS in Local Government

The pathways for GIS evolution, maturity and growth in local government are critical, game changing and dynamic in nature. *A town, city or county needs to know the exact maturity level of its GIS systems.* Today's approaches often lack clarity and fail to plan for the next stage of evolution, namely the "smart city" phase of GIS evolution in local government.

Every town, city and county is at a different stage of geospatial development and maturity. Some organizations are not ready to embrace new emerging technologies. Some organizations are moving ahead with the smart city concept. The speed at which organizations adopt geospatial technology is dependent on multiple factors.

I posit that the evolution, maturity and existing GIS competence levels in local government can be determined and documented using two specific methods of measurement. These are the six pillars of GIS sustainability and the six logical stages of GIS maturity.

Six Pillars of GIS Sustainability

Government organizations are often stymied by weaknesses in one (or all) of the six pillars of GIS sustainability. The six pillars of GIS sustainability include the following:

- Pillar One: **Governance** Components
- Pillar Two: **Digital Data and Databases** Components
- Pillar Three: **Procedures, Workflow and Interoperability** Components
- Pillar Four: **GIS Software** Components
- Pillar Five: **Training, Education and Knowledge Transfer** Components
- Pillar Six: **GIS IT Infrastructure** Components

(Please note: The term information technology (IT) infrastructure refers to the entire collection of hardware, software, networks, data centers, facilities, and related equipment used to develop, test, operate, monitor, manage and/ or support geospatial information technology services.)

Figure 1.15 depicts the six pillars of GIS sustainability.

FIGURE 1.15
Six Pillars of GIS Sustainability.

Six Logical Stages of GIS Maturity

Many analyses and descriptions of the ways GIS matures in local government already exist. Numerous GIS maturity models describe the different stages of growth. A simplified and rigidly linear system of a life-cycle approach to GIS maturity includes: a start-up stage, operational stage, stabilization stage, expansion stage and advanced stage. We can look at the maturing of GIS purely from the historic technology perspective. This approach includes project-based GIS, workgroup GIS, departmental-based GIS, enterprise GIS, networked and web-based GIS and wireless mobile-based GIS. The second perspective looks forward to the ubiquitous, embedded, transparent and interoperable "system of systems" that will define GIS in the near future. Thus, we can agree that there are numerous ways to describe the maturing of GIS in government. URISA has developed a maturity model that is detailed, is informative and includes stages of maturity.

I posit that there are six logical stages of GIS maturity from adoption to the smart city concept of tomorrow. The six logical stages of GIS maturity in local government are detailed in Figure 1.16.

We cannot evaluate and diagnose an organization using oversimplified models that fail to embrace the core components of GIS. These core components or building blocks of GIS are ingrained in the six pillars of GIS sustainability. Any life-cycle GIS maturity model must cross-reference these key components of GIS. How can we discuss GIS and geospatial growth and maturity without an in-depth analysis of the core components of GIS? My answer is: by looking at the evolution and maturity of GIS technology in local government organizations through a two-part lens synthesized from the six pillars of GIS sustainability and six logical stages of GIS maturity.

We can look at the combination of the six pillars of GIS sustainability and the six logical stages of GIS maturity in two distinct ways. One way is to look at the evolution and growth of GIS from each of the six pillars of sustainability. That is essentially asking what type of GIS governance, data and databases, procedures, workflow and integration (interoperability), GIS software, training and IT infrastructure an organization uses during its maturation? Another way of looking at the information is to look at each phase of maturity, from adoption to enhancement, operational to strategic and enterprise to smart and sustainable. Both ways are particularly useful for understanding GIS in local government.

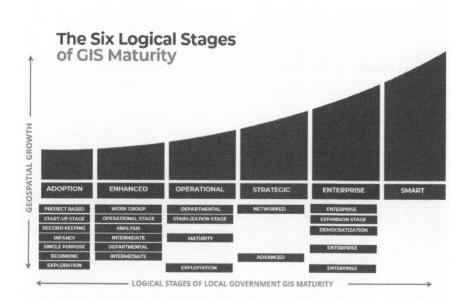

FIGURE 1.16
Six logical stages of GIS maturity.

It is important to note that fundamental strategic, technical, tactical, logistical and political challenges persist through each stage of GIS maturity. The six stages of maturity include:

Stage One: Adoption

The early adoption phase of GIS tends to be event driven. Generally, that means a single individual within an organization identifies the need for geospatial technology and map products. The adoption phase is characterized by the following:

- No GIS plan with ad hoc activity by one individual.
- A simple, singular database or rudimentary online data layers.
- Very little integrating or combining of data.
- A single desktop or cloud solution or free online application.
- A general awareness and understanding of the benefits of geospatial technology.
- Generally, an unsophisticated use of IT infrastructure. Desktop tools and access to the web.
- No outreach or engagement with residents.
- No governance model.

Stage Two: Enhanced Usage

The enhanced phase of GIS implementation is much like a thunderstorm gathering energy. It begins with a groundswell of enthusiasts. Though still isolated, the group of early adopters begins formulating ideas for innovative uses for GIS technology. They begin to see and articulate the benefits of GIS; however, their uses of the technology are project based, generally lacking coordination and distributed among a few individuals in specific departments. The enthusiasts only use limited spatial functionality to replace paper maps.

The enhanced phase of GIS is characterized by the following:

- No GIS plan or direction, with makeshift and informal group activity. It often includes one to five users within the organization.
- Multiple simple, individual silos of databases.

- Improvised and limited database integration.
- Multiple desktop and or cloud solutions in isolated operations.
- Growing GIS awareness, understanding and knowledge of geospatial technology.
- Departmental workgroups using IT infrastructure.
- No outreach or engagement with residents.
- No governance model.

Stage Three: Operational Efficiency

The operational efficiency or competency phase introduces new elements to a local government's relationship with GIS. These elements include departmental commitment, a broader understanding of the value of GIS technology, and a growth in the number of users requiring a GIS governance model.

The operational efficiency phase is characterized by the following:

- No GIS plan but deliberate, organized and arranged GIS activity by key individuals. Some action plans do exist to direct GIS activity.
- Multiple target-specific GIS silos of data.
- Collaborative and conceptual ideas about procedures, protocols and standards.
- Multiple workgroups, desktops, cloud solutions and business applications.
- Wider GIS awareness, understanding and knowledge and proficiency.
- Departmental workgroups using connected IT infrastructure – beginning of enterprise data storage.
- No real outreach or engagement with residents.
- No governance model but requires some governance.

Stage Four: Strategic Efficiency

The strategic phase is characterized by a deliberate move toward a true organizational GIS. It is formally or informally goal orientated. GIS is often introduced to comply with regulations. The chief characteristics of the strategic phase include:

- The existence of a GIS plan or roadmap that is structured, considered and goal orientated.
- GIS is integrated with structured departmental silos. It has the beginning of integration and interoperability.
- Incorporated, aligned and task-specific GIS procedures, protocols and standards.
- Targeted groups of desktop, online, mobile and business applications.
- Formally trained and educated users. Thoughtful and degreed. Classroom and online webinars, seminars and GIS certification.
- Networked, web enabled and enterprise data storage.
- Geospatial technology for outreach or engagement with residents.
- Centralized with some decentralized characteristics GIS governance model.

Strategic implementation is characterized by methodologies that are deliberate, considered, intentional and tactical. It is supported by well-calculated planning and decision making. Strategy means examining the big picture – keeping long-term goals and objectives in mind while carefully analyzing the actions and initiatives required to reach those goals. It can include a vision, goals and objectives supported by key performance indicators (KPI) and GIS outcomes.

Stage Five: Enterprise, Strategic and Operational Excellence

The phase of enterprise, strategic and operational excellence is characterized by an organization-wide desire to plan, design, implement and maintain a corporate GIS. There is a department- and organization-wide commitment to GIS and an advanced understanding of the benefits of the technology. In this phase, GIS is used routinely by the majority of departments. This stage includes the following characteristics:

- A GIS strategic enterprise plan that is tactical, technical, logistical and political. It normally includes a cost-benefit or ROI analysis and buy-in from senior staff.
- Integrated and embedded with bi-directional functionality.
- Managed and sustainable GIS procedures, protocols and standards.
- Extensive desktop, online, mobile, business application, community and crowdsourcing tools, open data, GIS task-specific extension, bespoke GIS solutions and advanced analytics capabilities.

- Formally trained and educated users. Experienced, certified, qualified and accredited.
- Enterprise-wide IT infrastructure with managed applications, hosting, servers, storage and backup, networked, web enabled and enterprise data storage.
- Geospatial technology is used extensively for outreach or engagement with residents.
- A hybrid governance model is deployed to organize a core central team of experts that enables the decentralized users.

Stage Five has many moving parts. These include planning, designing and implementing GIS solutions throughout entire organizations. All departments deploy geospatial technology creatively and originally. The staff is willing to undertake a significant project or task ... especially if the task is complicated, difficult or risky. Enterprise is associated with boldness, resourcefulness and energy.

Stage Six: Smart and Sustainable – Smart City

This phase of GIS implementation is the most advanced phase we have today. The smart and sustainable phase, or "smart city" phase, includes community interfacing, citizen feedback, open data, story maps and extensive participatory GIS. This stage is characterized by the following:

- A strategic, enterprise-wide, scalable, sustainable, enduring and smart GIS business plan that includes GIS benefit realization planning (quantifying and qualifying the benefits of GIS), high-performance organizations (HPO) strategy (a new way of managing the organization) and a scorecard of sustainability and resilience.
- Succession planning is addressed.
- Total interoperability and bi-directional functionality. A "system of systems" with real-time data.
- Implemented and maintained state of the art GIS procedures, protocols and standards. Data standards are a key factor in this stage.
- Total GIS ecosystem with big data analytics.
- Certified and trained users. Innovative, imaginative, transparent, original and pioneering. An organizational curriculum.

- Enterprise-wide IT infrastructure with managed applications, hosting, servers, storage and backup, networked, web enabled and enterprise data storage.
- Geospatial technology used extensively for outreach, engagement, interaction and feedback from residents.
- A hybrid and regionalized governance model is used.

Figure 1.17 depicts the six logical stages of GIS maturity and geospatial growth in local government.

A truly sustainable GIS is resilient. It can handle the weight of economic downturns, key staff changes, technological shifts, political imbalance and innovative new solutions. Sustainability means to progress or continue; to nourish something into long-term fruition. Sustainability requires intentionality, support and maintenance. A scalable GIS must be flexible and deployable in many different ways that including intranet, internet, desktop and mobile GIS technologies. Scalability also involves customization – or the potential for the GIS systems to be upgraded, expounded upon or simplified to fulfill a task. It hinges on accommodation and cooperation. As such, enterprise-wide, sustainable solutions are scalable to all needs of government. An enduring GIS solution is long-lasting, permanent and withstanding. Endurance requires an organization to work at something diligently, sustainably and meticulously in order to produce lasting outcomes that remain constant and stable for long periods of time.

The next question we need to address is twofold: First, how is GIS evolving or implemented within local government? Second, how do we evaluate or diagnose its success and maturity while accounting for setbacks, challenges, barriers and pitfalls? It is absolutely vital that we understand the ways GIS matures in local government and how that process relates to the core components of GIS technology. This understanding is fundamental to implementing future smart practices and applications. An organization that has matured into a true strategic, enterprise-wide, operationally efficient, sustainable, scalable, enduring and smart GIS is absolutely doing something right. In all likelihood, such an organization has an extensive portfolio of smart practices and applications. As GIS practitioners, we must be cognizant of how GIS grows and matures within local government organizations.

Further Questions?

We should ask: why is it important for an organization to know what stage of maturity it is in? How do local government organizations recognize and measure evolutionary "stages"? Can we even agree what constitutes a GIS maturity stage, much less whether an organization has entered one? Does a local government have to go through each stage of maturity, or can they skip

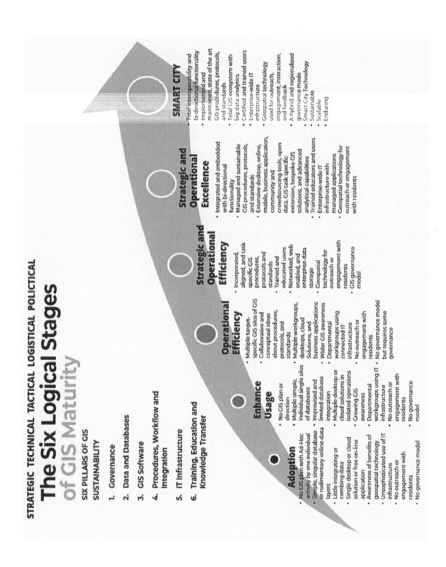

FIGURE 1.17
Six logical stages of GIS maturity and geospatial growth in local government.

stages? How fast can an organization move through the stages? Can they get through them faster? Should they try?

Well, since the 1960s, local governments have been shaping GIS as we know it! Though not a concrete answer, it is worth noting that the history of municipal GIS is filled with distinct evolutionary changes in geospatial technology. That brings us to another question: what key factors have influenced or are influencing GIS today? To begin answering these questions, we must consider what the academic world calls "paradigm shifts" in the GIS landscape.

Geospatial Paradigm Shift or Natural Evolution

Geospatial technology paradigm shifts differ fundamentally from natural, albeit accelerated technological evolutions. This distinction is worth considering. Though there are many forces shaping the future, it is the organic geospatial technological advancement that has occurred since 1960 that proves most crucial. Today, however, we are witnessing a cultural change in the character of our geospatial world. Does this 2020 transition to ubiquitous participatory GIS count as a paradigm shift?

"Paradigm shift" refers to a new mode of doing or understanding something that is profoundly different than the prior mode. The term "paradigm shift" could have entered our vocabulary for the express purpose of describing revolutionary changes in technology. Paradigm shifts mark radical departures from the previous era's way of doing things.

In an article entitled "Geospatial Paradigm Shift or Not?" (re-published in GIS Café), writer Carl Reed cites Thomas S. Kuhn's book *The Structure of Scientific Revolutions*. As Kuhn states, a profound scientific breakthrough or paradigm shift "is seldom or never just an increment to what is already known. Its assimilation requires the reconstruction of prior theory and the re-evaluation of prior fact, an intrinsically revolutionary process that is seldom completed by a single person and never overnight." Kuhn referred to these revolutionary processes in science as "paradigm shifts."

> Example: When Ferdinand Magellan's practical demonstrations challenged the widely held assumption that the Earth was flat, someone could have fairly said "I think we have a paradigm shift on our hands."

In the geospatial world, we have seen a number of significant "increments to what is already known." We have often built upon but not reconstructed a prior theory. Remember from the beginning of this chapter that **the future is always the present. We react, we do not predict.** The same goes for technological shifts. Over the last thirty years, we have taken existing ideas and

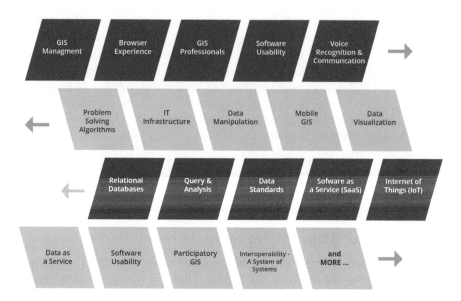

FIGURE 1.18
Geospatial technology shifts.

prior theories and improved upon them. We did not re-evaluate prior facts nor did we reconstruct prior theories.

Geospatial technology grew remarkably over the last forty years. This process included both rapid organic evolutions and significant technological shifts. Let us briefly review the changes in geospatial technology in local government since the 1960s. Note that it's always difficult to separate geospatial technology advancements from advancements in other related and co-dependent technologies. We have seen a massive shift in the ability of technologies to talk to each other. We are now losing the term "GIS" in favor of "geospatial technologies." This is important to recognize as we are approaching a significant shift in how we think about, manage and apply GIS technology. Let's look at what we call "geospatial shifts." Figure 1.18 lists some important technology shifts in geospatial science.

New Governance, Philosophies, Values, Beliefs and Attitude Shifts: I reviewed one of the first 2019 requests for information (RFI) from a municipality advertising for outside help with the smart city concept. The City of St. Louis Smart City Advisory Group sought information regarding smart city technology solutions for the following list of goals. Figure 1.19 lists the City of St. Louis' smart city goals (City of St. Louis).

- **Public Safety** – Make the City safer for residents and visitors.
- **Digital Equity** – Increase access to technology for all citizens of the City.

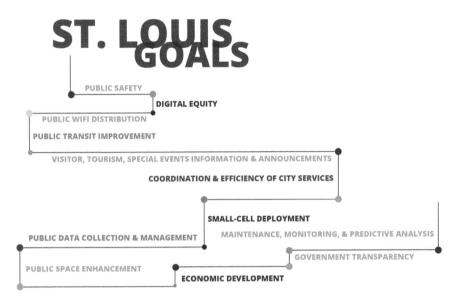

FIGURE 1.19
City of St. Louis smart city goals.

- **Public Wi-Fi Distribution** – Make Wi-Fi internet access available throughout the City.
- **Public Transit Improvement** – Make public transportation simpler and safer in order to grow the City's transit ridership.
- **Visitor, Tourism, Special Events Information and Announcements** – More efficiently disperse information to the people who live, work and explore the City.
- **Coordination and Efficiency of City Services** – This includes but is not limited to: maximizing energy and cost savings and decreasing service delivery timeframes.
- **Small-Cell Deployment** – Manage the deployment of small-cell technology in a coordinated effort that will both make the technology available and keep a common theme throughout the City.
- **Public Data Collection and Management** – Collect data about how people move through and enjoy the City in order to make the City more efficient and accessible and enhance data sharing and collaboration between City Departments and other agencies.
- **Public Space Enhancement** – Improve the look and feel of the City by reducing streetscape clutter and improving lighting.
- **Economic Development** – Drive investment, development and population growth in the City.

- **Government Transparency** – Increase government transparency and community engagement.
- **Maintenance, Monitoring and Predictive Analysis** – Establish processes for better management of City infrastructure.

Which goal doesn't require geospatial technology? The answer is: none. All of these goals lend themselves to the world of GIS and geospatial technology. This RFI is either a testament to the constant pressure for municipalities to do more with less or further affirmation that the world of local government is poised – yet again – to precipitate some of the most innovative and interesting applications of geospatial science. With their desired technology and application shifts, the City of St. Louis was really asking for a fully geo-smart government!

Digital Data and Databases Shifts: Advances in technology, procedures and protocols are improving techniques for data collection, generation and analytics. These overarching trends create new data expectations, new visions for data sharing, the demand for open and transparent data, social media data, seamless business integration, improved data access and delivery in addition to new ways of analyzing the relationship between data and the improved geospatial thinking and reasoning that it affords. The question for a local government is: what is your vision and expectation for data and databases?

Fundamental smart practices must be in place for all this advanced data analysis to occur. This is the primary issues we must address. Figure 1.20 reminds us that data can be fickle. We must constantly assess the content and quality of data.

Technology Shifts: New and emerging technology trends and worldwide smart city (a systems of systems) initiatives, including real-time data, inexpensive miniature devices and sensors, the Internet of Things (IoT), Unmanned Aircraft Systems (UAS) and drone technology, mobile solutions, expanding wireless and web networks, advances in computer capacity and speed, data collection and data generation, big data analytics, data models, improved communication tools and new software applications are all parts of the future landscape. With all of those factors in mind, we have to ask: what are the mission critical technological priorities for a local government organization? We are in the middle of a geospatial technology revolution that is transforming our governments, citizens, economy and environment. Smart GIS practices in local government are joined at the hip with the technological and political landscape of the day. The **Smart City** concept refers to a **system of systems that combines innovative technologies and applications – the Internet of Things (IoT), real-time data, Data collection devices, crowdsourcing, visualization, configurable web apps, 3D, social media, GPS, remote sensing, LIDAR, virtualization, faster computing, cloud computing and consumerization.**

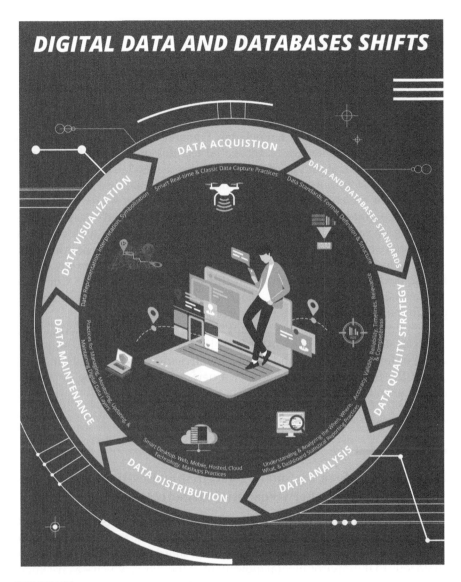

FIGURE 1.20
Data and data assessments.

Civic Engagement and Management Shifts: GIS-based public or civic engagement applications (apps) improve the performance and image of local government by helping residents and stakeholders actively participate in governmental operations on their own terms. Civic management requires an examination of the ways that geospatial technology – including usable crowdsourcing solutions, external agency interfaces, bi-directional applications and open data portals with instant real-time feedback mechanisms – positively affects a local government. How does this new shift toward participatory GIS, public engagement, open data portals and crowdsourcing solutions fit into the future of local government? Already, the practice of participatory GIS is engineered for community empowerment through user-friendly, measured, demand-driven and integrated cloud-based geospatial solutions. It goes without saying these practices will continue evolving in the future!

Training, Education and Knowledge Transfer Shifts: A workforce fully prepared for the challenges of geospatial technology is crucial to the success of any organization. An optimum governance model with clear lines of responsibility, communication and structured accountability and workforce management are characteristics of a true enterprise geospatial solution. A governance strategy that facilitates the ongoing research and development of emerging geospatial technologies provides an organization with resilience and sustainability. Defining the critical components of a true enterprise and sustainable geospatial governance model will challenge GIS professionals for the next twenty years. How will this model promote better workforce management and engagement? What role does research and development play in the sustainability of the organization?

Well, first of all, we will see a "shift" in the way local governments train and educate their staff. The City of Berkeley just completed a new and innovative online GIS training and education curriculum for City staff. It marks one of the first attempts to engineer a new way of supporting geospatial initiatives in local government.

What Do We Mean by Smart, Sustainable, Resilient and High-Performance Geospatial Organizations?

The future of GIS and geospatial technology will align itself with smart, resilient and sustainable communities in addition to the development of HPO. We could venture to say that the recent smart city and HPO events are analogous to the environmental movement of the 1960s, which – as we know today – catalyzed significant changes in the history of GIS. The pollution of the Great Lakes in the 1960s became a "rallying point for environmentalism." Oil spills, lake fires and other natural and human-made disasters promoted an increased interest in the preservation of America's natural landscape. The environmental movement of the 1960s and 1970s spurred the development of environmental science as a discipline that provided new tools to collect, view, analyze and model data. The reality of today's smart technology, remote sensors, advanced analytics and the potential held by tomorrow's technological GIS ecosystem or "system of systems," begs the questions: are we at the beginning of something revolutionary for the world of geospatial science in government or are these developments just a natural evolution walking in step with other technological advancements?

What Is a Smart Organization?

There is no universally accepted definition of a smart city or smart organization. It means different things to different people. The smart community concept varies from city to city, town to town and country to county. The concept itself often depends on receptivity to change and reform at the local level. **Energy, water, waste and air** are all part of a smart city initiative. The smart city's objective is to cultivate community infrastructure, give decent quality of life to its citizens, nurture clean and sustainable environments and apply "smart" solutions. The focus is on sustainable, resilient and inclusive development.

The following excerpt was taken from the City of Mississauga's Smart City initiative. "Mississauga is committed to engaging the community, citizens and industry leaders in our Smart City journey. A Smart City is a municipality that uses data and communication technologies to create sustainable economic development, increase operational efficiency, improve the quality of government services and make improvements to community life." The City goes on to explain that it has been "recognized for its advancements in Smart City initiatives including free public Wi-Fi, open data, hackathons, fiber network, Advance Traffic Management (ATM), LED lighting, mobility, online services, mobile apps and social media. In addition, Mississauga has connected City services such as building automation, snow ploughs, the works fleet, transit and fire." By deploying the following smart technologies, the City of Mississauga fashioned itself into a true twenty-first century city.

1. **Creating a Fiber-Connected City**: The City of Mississauga is home to the largest municipally owned fiber optic network in Canada. Over 800 kilometers of fiber connect over 290+ sites across the city.

2. **Fostering Digital Inclusion**: The City offers free wireless internet access in community facilitates and public spaces, with more locations under review.

3. **Supporting Innovation**: The City of Mississauga currently has over 100 data sets available on their Open Data (OD) catalogue.

4. **A Virtual City Campus**: The City of Mississauga is the first Canadian city to make Eduroam available to its student population, creating a virtual campus using the City's free public Wi-Fi network. Please note that "Eduroam" is an "international roaming service for users in research, higher education and further education. It provides researchers, teachers and students easy and secure network access when visiting an institution other than their own" (Google).

5. **Engaging Youth and Tech Community**: The City is dedicated to engaging youth and building a strong, future-forward community by supporting technology events.
6. **Mobile and Digital Transformation**: The City prides itself on their online service deliveries such as self-service taxes, license renewals and recreation program registration.

The City of Mississauga will

> "continue to invest in innovation and new ways to use data and technology as we build a modern and progressive city for the 21st century. (They) will continue to build a smart city of vibrant communities where everyone has equal opportunity and feels empowered, a place where people can connect, adapt and succeed"

(Mayor Bonnie Crombie, City of Mississauga).

The concept of smart cities or smart organizations is about designing sustainable urban environments for the modern world. The objective of a **smart organization** is to develop a connected technology ecosystem to improve the quality of life for citizens. **Smart technology** harnesses digital data from smart devices including the Internet of Things (IoT), networks, cloud infrastructure and software applications. The end product of all this data comes in the form of analytics that provide new insights, products and services. The **smart platform** denotes a technology allowing for the integration and bi-directional interoperability of smart technologies. The smart platform of a connected enterprise ecosystem effectively and seamlessly collects, combines and manages digital data, as well as enabling new software applications.

The slow transformation of organizations around the world is underway with the increased use of embedded smart devices and IoT. These embedded devices are connected to a network that becomes part of a system, which is in turn one cog in the "system of systems" that encompasses an entire organization. So what does a local government organization need to recognize about the required outcomes of a smart initiative?

1. Allowing **personalized recognition between people and systems** (example: Bike and scooter sharing)
2. **Providing location-based information** (on-demand buses, taxis, trains, ferries and more)
3. **Deploying geosensors** – observing, understanding and anticipating events (air quality, noise and advanced climate and weather information)
4. **Linking people** to the community and to services, resources, amenities and one another (traffic counts, Tweets)
5. **Interoperability** – enabling different systems, information sources and data types to work together (CAD, GIS and BIM)

The large and complex municipal organizations governing places such as Singapore, Barcelona, London, San Francisco and Oslo are all committed to the smart city concept. These organizations are using smart technology to manage their urban areas.

Canada already holds a "smart city challenge." This challenge is a pan-Canadian competition open to communities of all sizes, including municipalities, regional governments and indigenous communities (First Nations, Métis and Inuit). The challenge encourages communities to adopt a smart cities approach to improve the lives of their residents through innovation, data, and connected technology. Canada has allotted over $75 million in prize money.

- One prize of up to $50 million open to all communities, regardless of population.
- Two prizes of up to $10 million open to all communities with populations under 500,000 people.
- One prize of up to $5 million open to all communities with populations under 30,000 people.

The Canada Challenge Statement allows each Canadian community to define its "challenge statement." The challenge statement is a single sentence that defines the outcome or outcomes a community aims to achieve by implementing its smart city proposal. The challenge statement must be measurable, ambitious, and achievable through the proposed use of data and connected technology (Infrastructure Canada, 2019). Examples of challenge statements are:

1. **Feel safe and secure**

   A neighborhood in our community with the highest crime rate will become safer than the national average.

2. **Earn a good living**

   Transform a former industrial neighborhood into one of the top locations in Canada for economic growth.

3. **Move around my community**

   Ensure that every senior who is able to live independently at home is empowered to do so.

4. **Enjoy a healthy environment**

   Implement preventative measures to reduce flood damage risk by forty percent and provide every resident of at-risk areas with access to these measures.

5. **Be empowered and included in society**

   Ensure that every person without a home has access to nightly shelter, and will connect one hundred percent of vulnerable residents with the services, activities and programs that are known to reduce the risk of homelessness.

6. **Live an active and healthy life**

   Enable the population to become fifty percent more active and healthy and thus achieve measurable decreases in chronic disease.

What Are Sustainable Communities?

Sustainable communities focus on environmental and economic sustainability, urban infrastructure, social equity and municipal government. The term is synonymous with the phrases "green cities," "eco-communities," "livable cities" and "sustainable cities."

Sustainable communities can be described as places where the needs of everyone in the community are met and people feel safe, healthy and ultimately happy. The environment is appreciated, protected and enhanced. Sustainability focuses on the following technologies:

- Wind energy
- Solar energy
- Sustainable construction
- Efficient water fixtures
- Green space
- Sustainable forestry

One of the best descriptions of sustainably communities came from Mr. Kaid Benfield, director of the Sustainable Communities and Smart Growth program at the Natural Resources Defense Council, when he said

> Sustainable communities are places where per capita use of resources and per capita emissions of greenhouse gases and other pollutants are going down, not up; where the air and waterways are accessible and clean; where land is used efficiently, and shared parks and public spaces are plentiful and easily visited; where people of different ages, income levels, and cultural backgrounds share environmental, social, and cultural benefits equally; where many needs of daily life can be met within a 20-minute walk, and all may be met within a 20-minute transit ride; where industry and economic opportunity emphasize healthy, environmentally sound practices.

Benfield, 2011

What Is a Resilient Community?

A resilient city is one that has developed the capacity to absorb **future shocks** to its social, economic, and technical systems and infrastructure. A resilient community can maintain essentially the same functions, structures, systems and identity in the event of a crisis. A resilient community manages the risk of disasters such as:

- Climate change
- Economic crisis
- Health epidemics
- Uncontrolled urbanization

Organizations assess their strengths, vulnerabilities and exposure to natural and manmade threats and then develop strategies and solutions accordingly.

What Is a High-Performance Geospatial Organization?

Managing services that affect the welfare of all citizens and residents of a community is a complex challenge. Local governments are charged with providing vital community services such as police, fire, emergency services, water, wastewater, refuse collection, building development, housing services and social services.

Over the last three decades, technological change has been a unifying constant for local governments across the world. The need to evolve and reform the way a city, town or county delivers geospatial technology internally and externally will continue to challenge government and GIS professionals. Additionally, geospatial technology success in government will remain dependent upon local governmental leaders and their understanding of the future's geospatial environment. As Kenneth Cloke and Joan Goldsmith (2002) stated in their book, *The End of Management and the Rise of Organizational Democracy,* "Every organization requires administration, coordination, facilitation, and leadership," and the organization of the future should be "an organization that must be capable of producing high-quality, competitive products that satisfy customers without destroying the planet or degrading human life."

We should quickly review the historic timeline of management and leadership styles. In broad strokes, we moved from the Roman Empire's reliance on slavery to the widespread medieval institution of "serfdom" that remained utterly unconcerned with the rights of workers, all the way to Frederick

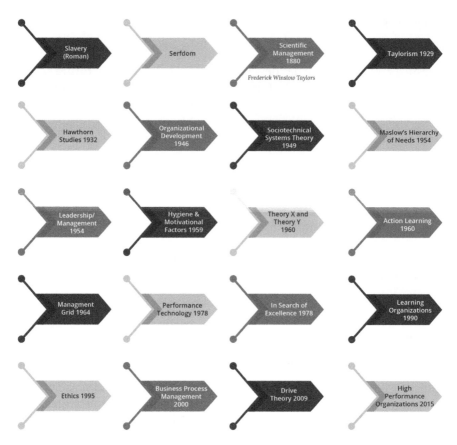

FIGURE 1.21
History of management and leadership.

Winslow Taylor's modern management strategy. Taylor was a nineteenth-century American engineer known for improving economic efficiency and developing his management theory known, appropriately, as Taylorism. Figure 1.21 illustrates the changing nature of management and management styles since the Roman Empire.

- Slavery (Roman)
- Serfdom
- Scientific Management 1880 – Frederick Winslow Taylors
- Taylorism 1929
- Hawthorn Studies 1932
- Organizational Development 1946
- Sociotechnical Systems Theory 1949

- Maslow's Hierarchy of Needs 1954
- Leadership/Management 1954
- Hygiene and Motivational Factors 1959
- Theory X and Theory Y 1960
- Action Learning 1960
- Management Grid 1964
- Performance Technology 1978
- In Search of Excellence 1978
- Learning Organizations 1990
- Ethics 1995
- Business Process Management 2000
- Drive Theory 2009
- High Performance Organizations 2015

I am confident that each stage of social development outlined in Figure 1.21 contributed significantly to the current theory of management. Take, for instance, Frederick Taylor's 1880 statement that "In the past man was first. In the future the system will be first." (Taylor, 1911). What could be more prescient?

Today, we need to look at how the HPO management strategy can be applied to geospatial technology in local government. The **HPO** is a conceptual and scientifically validated structure that managers use when deciding what to focus upon in order to improve and sustain organizational performance. The HPO framework, when translated by managers to their specific organization, is a recipe for success. The framework includes the characteristics listed in Figure 1.22.

The key HPO questions that determine success include:

- Who are currently our stakeholders (beneficiaries, customers, partners, collaborators, suppliers and others)?
- Who should be our stakeholders (potential stakeholders)?
- What do each of our stakeholders or potential stakeholder's value (wants/needs/expectations)?
- What does high performance mean to us?
- How does our vision "nest" within that of the organization?
- How would we describe the desired future state we are seeking?
- How would we know if we were high performing (metrics)?
- How do we define quality for our products and services?
- How are we going to treat each other and our stakeholders?

FIGURE 1.22
Characteristics of a high-performance organization.

HPO is predicated upon an organization's continuous improvement and can be defined by the following six parameters:

1. **Improvement based on small changes**: Improvements that are based on minute organic changes rather than the sweeping changes that arise from research and development.
2. **Staff ideas increase the rate of adoption**: As the ideas come from the workers themselves, they are less likely to be radically different, and therefore easier to implement.
3. **Small improvements are cost effective**: Small improvements are less likely to require major capital investment than major process changes.
4. **Working knowledge and talent**: The ideas come from the existing workforce as opposed to researchers, consultants or new equipment – any of which could prove very expensive.
5. **Involves everyone in the organization**: All employees continually seek ways to improve their own performance.

6. **Staff take ownership**: Taking ownership of results encourages staff and reinforces collaboration, thereby improving worker motivation.

What Is Business Realization Planning?

It remains difficult for local government organizations to identify, measure and monitor the performance of geospatial operational business functions in both quantitative and qualitative terms.

Historically, the GIS cost-benefit analysis (CBA), value proposition and/ or the geospatial ROI analysis have been deployed by local governments to assess the value of geospatial technology. The Business Realization Plan (BRP) includes the business case, appropriate measures, benefit drivers, processes and the ongoing monitoring of geospatial technology's benefit to the organization. This is a new and comprehensive mode of evaluating of geospatial technology in local government.

The BRP is a way to measure how projects and programs add true value to the organizations. It provides a set of questions and good practices that local government management professionals and leaders can use to guide the identification, analysis, delivery and sustainment of benefits aligning with the organization's strategic goals and objectives.

The Project Management Institute (PMI) Thought Leadership (2016) series outlines the following strategies as components of business realization planning:

A. Identification of geospatial benefits to determine whether projects and programs can produce the intended business results.

B. Execution of geospatial benefits to minimize risks to future benefits and maximize the opportunity to gain additional benefits.

C. Sustainability of geospatial benefits to ensure that whatever the project or program produces continues to create value.

Figure 1.23 illustrates the key components of business realization planning.

GEO-SMART GOVERNMENT: Smart Practices and the Future Smart City. Deconstructing Our Complex Geospatial Future: An Altogether Different Language
Researchers have extensively documented the unprecedented migration of the world's population from rural to urban areas over the last century. Widespread urbanization continues to precipitate whole new sets of social and infrastructural challenges for cities. Take the words of Dorothy

KEY COMPONENTS
OF BUSINESS REALIZATION PLANNING

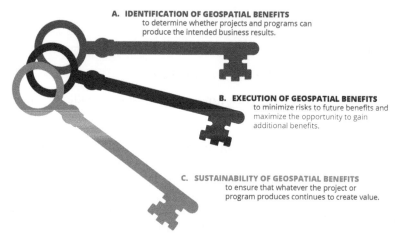

A. IDENTIFICATION OF GEOSPATIAL BENEFITS
to determine whether projects and programs can
produce the intended business results.

B. EXECUTION OF GEOSPATIAL BENEFITS
to minimize risks to future benefits and
maximize the opportunity to gain
additional benefits.

C. SUSTAINABILITY OF GEOSPATIAL BENEFITS
to ensure that whatever the project or
program produces continues to create value.

FIGURE 1.23
Key components of business realization planning.

Parker, "Los Angeles is 72 suburbs in search of a city." This mass movement of people to urban areas necessitates that government officials, city planners, businesses and residents embrace new ways of managing people and spaces. New technologies, processes, products and engineering solutions are constantly being created to address the unique issues resulting from this unprecedented rate and scale of urbanization.

According to some researchers, in one hundred years humanity will be an entirely urban species. British-Canadian author and journalist, Doug Saunders visited thirty villages and cities on five continents to explore the great migration from the countryside to vast megacities. This movement is a singular challenge for humankind. Saunders writes, "Major technological, economic, and environmental changes are generating interest in smart cites, including climate change, economic restructuring, the move to on-line retail, and entertainment, aging populations, urban population growth" (Saunders, 2010).

The past is the most important reference point for determining what will happen in the future. Before you read the next section, try to remember what I have hashed and re-hashed in this chapter: that we do not really predict the future. We deploy existing tools and ideas in better and more original ways. We guess and calibrate according to our understanding of how large these newly re-engineered applications of geospatial science will prove to be. **The future is always the present. We react, we do not predict.**

Smart Practices and the Future Smart City GIS

The phrase "Best Business Practices" (BBP) gets thrown around the corridors of local government like confetti. But what are BBP? Let's assume for the sake of argument that it is a method, technique or procedure that has been accepted and embraced as the best way of doing something. It's the superior alternative to other procedures, a standard way of doing things or a template that's agreed upon as the best blueprint for accomplishing a specific task.

The problem with the word "best" is that it's entirely subjective. There is no world organization like the Guinness World Records (GWR) that keeps a database of all known, agreed upon and accepted BBPs. Alas, the GWR's aim to "make the amazing official" and be the "ultimate global authority of record keeping" has no analogue in the world of BBP. Like a world record, the shelf life of BBPs can vary greatly. Unlike a world record, a BBP is never officially the world's best. It's simply accepted that a given method is or was a BBP. At best, BBP are working solutions that are labeled "best" by the arbiters of conventional wisdom.

What about using the term "smart practices" instead? Perhaps, if a method, technique or procedure worked exceptionally well within an organization, it simply becomes known as a "smart practice?" Instead of it being "the best," a practice could simply be a smart practice, a good practice or an innovative and promising practice. This term frees us up to consider a variety of smart practices. That's why I employ the modest, practical and appropriately encompassing term "smart" geospatial practices as it relates to local government. Thus was born the title of this book *Smart GIS Practices and Applications in Local Government: An Altogether Different Language*. Now, let's move on to another original term: geo-smart government.

Defining the Smart City or a Geo-smart Government

Think about your air conditioning unit at home. You've set the thermostat to a specific temperature. As the temperature changes, the heat sensor in the thermostat picks up the change and engages the air conditioning unit to turn on or switch off. This is a smart device. It's conceptually the same for all other smart devices. Smart devices and sensor technology are automating and advancing traditionally manual processes. We have smart refrigerators, smart ovens, smart toothbrushes and now, smart cities. Smart cities use many types of electronic data to manage assets and resources. Data is collected from an assortment of field devices and residents.

The exact definition of a smart city depends on what you read and who you talk to. The ultimate goal of a smart city is to optimize functionality and drive economic growth while improving the quality of life for residents through the use of smart technology and data analysis. Theoretically, any area of city management can be incorporated into a smart city initiative. Let's discuss the intersection of smart and geospatial technologies plus their implications for local government. Figure 1.24 illustrates the long list of smart city initiatives.

From 2020 onwards, we will see great changes. It's not just technological changes we need to be mindful of. It's something much bigger, something in the hearts and minds of the next generation. It's a philosophy, an idea. It's the democratization of geospatial technology. It's the desire to make this technology understandable, available, participatory and ever more socially equitable. The cover of Harvard Business Review July–August 2019 states "The AI-Powered Organization: The Main Challenge isn't technology. It's Culture."

There is no doubt that every local government will continue incorporating geospatial technology into its strategic and operational activities. This will include deep learning, artificial intelligence, a new approach to the content and data … but how to use all this data? How to make data more useful and actionable? How to integrate machine learning, predictive modeling, big data analytics, 3D modeling, sensor fusion, building information modeling (BIM), augmented reality, virtual reality and mixed reality and data-based decision making across a variety of technologies and platforms?

They say, "If you want to know the future listen to what people are saying." What we are witnessing is an overall shift in the language of GIS. Ann Porter wrote *An Altogether Different Language*.

An Altogether Different Language by Anne Porter

> There was a church in Umbria, Little Portion,
> Already old eight hundred years ago.
> It was abandoned and in disrepair
> But it was called St. Mary of the Angels
> For it was known to be the haunt of angels,
> Often at night the country people
> Could hear them singing there.
> What was it like, to listen to the angels?
> To hear those mountain-fresh, those simple voices
> Poured out on the bare stones of Little Portion
> In hymns of joy?
> No one had told us.
> Perhaps it needs another language
> That we have still to learn,
> An altogether different language.

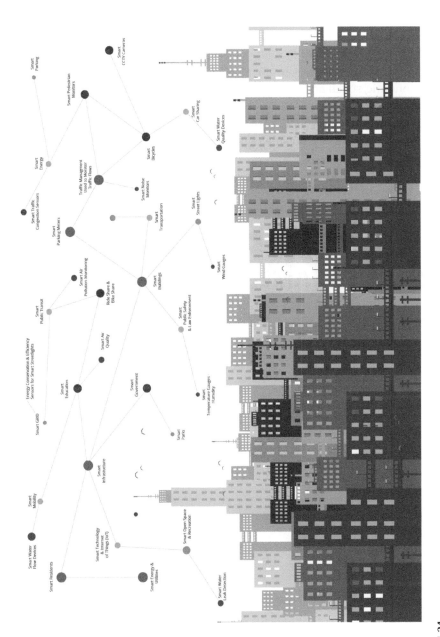

FIGURE 1.24
Smart city initiatives.

Our "altogether different language" is that of geo-smart government technology, culture, performance and value. Perhaps this new language would include the following:

- Smart city **GEOSPATIAL TECHNOLOGY** is a connected technology ecosystem. It is a system of systems.
- Sustainable and resilient **CULTURE** is the web of environmental, economic, social and technical conditions required for sustainability and resilience.
- High-**PERFORMANCE** organizations predicated on organic growth and continuous management improvement.
- Business Realization Planning and the **VALUE** proposition, or measuring key performance and business functions.

Figure 1.25 depicts the future of GIS management.

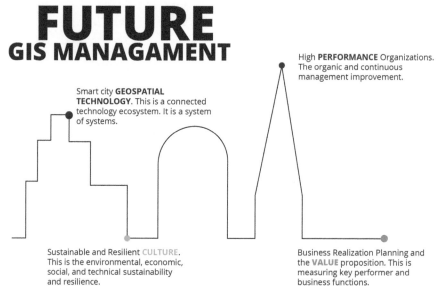

FIGURE 1.25
Future GIS management.

2

Smart Geospatial Technology Standard Practices, Activities, Procedures and Protocols in Local Government

It's supposed to be automatic, but actually you have to push this button.

– John Brunner

Introduction

Regardless of their evolutionary stage, local government geospatial initiatives require smart industry standard practices, activities, procedures and protocols. Each evolutionary stage of geographic information science (GIS) **requires planning, design, implementation, maintenance** and **reliable and repeatable practices to guarantee continued success.** These stages include but are not limited to:

- Stage 1: GIS Adoption
- Stage 2: Enhanced GIS Usage
- Stage 3: GIS Operational Efficiency
- Stage 4: Strategic GIS Utilization
- Stage 5: Enterprise GIS Usage
- Stage 6: Smart and Sustainable GIS

Each evolutionary stage grows geometrically from the previous stage. It is difficult for a municipality to go from Stage 1: GIS Adoption to Stage 5: Enterprise GIS Usage, overnight. The ultimate success of each evolutionary GIS stage is predicated on a continued advancement of standard practices, activities, procedures and protocols. Smart geospatial practices and application are therefore critical to local government success.

The geospatial technology of the future will only increase in complexity. We already see different types of map projections, different levels of location accuracy and many different types of GIS users. There are myriad different

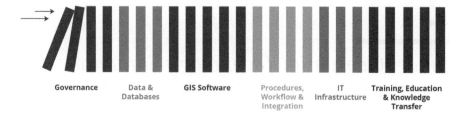

| Governance | Data & Databases | GIS Software | Procedures, Workflow & Integration | IT Infrastructure | Training, Education & Knowledge Transfer |

FIGURE 2.1
Knocking down all the smart practices, activities and procedures and protocols dominos.

ways to design geospatial databases, portray map information and collect data. Additionally, we have an abundance of software tools and many methods of encouraging staff.

The sheer number of methods and approaches in the geospatial technology landscape make the consideration of a uniform and accepted approach more important than we may think. I am working with a number of organizations that have failed to develop smart standards, procedures or protocols for their geospatial initiative. An ad hoc approach to GIS implementation is a recipe for disaster. In fact, these organizations are in the process of developing plans to re-engineer their program, starting with the critical long-term smart practices, activities, procedure and protocols.

The larger challenge for these and other organizations is to ratify, enforce, monitor, maintain and sustain good GIS practices. Essentially, you have to **knock down all the "practices and procedural" dominos before you see sustained success** (see Figure 2.1). The art comes in understanding that every organization must set up many "dominos" that come in the form of smart practices. Once you begin the process, all of your dominos will begin falling in line.

During the development of this section, I referred to my 2017 book *GIS Strategic Planning and Management in Local Government* (Holdstock, 2017), specifically the section that discusses the crucial factors for GIS success in local government. Moreover, I determined that this section needed some wisdom from local government GIS managers actually working in local government today … so I sent a draft off to a number of GIS professionals to acquire some feedback.

Understanding Smart Practices, Activities, Protocols and Procedures

Before we embark on a chapter about smart GIS practices, activities, protocols and procedures, let me explain what I mean by these terms.

GIS Standard Practices: These are the actual *applications or uses of an idea, belief or method*, as opposed to an examination of relating theories. They are the standard practices for an organization and require regularly updating GIS data layers and creating metadata standard forms for consumption by government staff and residents.

GIS Activity: This is the *behavior or action* of a particular kind. The act of publishing digital GIS data and populating metadata forms to be used in an open data portal for consumption by stakeholders is an example of an actual activity performed by a GIS analyst, manager or coordinator or even a geographic information officer (GIO).

GIS Procedures and Protocols: This is the fixed, *step-by-step sequence of activities, procedural methodology* or *course of action* that must be followed in a sequential order to correctly perform a task. A detailed step-by-step sequence of software tasks that are required to successfully share data and information across an enterprise is an example of procedures and protocols for data sharing. Repetitive **procedures** are called **routines**.

Latest Newfangled, Innovative and Novel Local Government Geosmart Activity That Might Require Formally Documented Smart Practices, Activities, Protocols and Procedures

Before we detail all the smart practices, activities, and procedures and protocols that a local government organization should consider before embarking on a GIS journey, we should make a mental note of all those up-and-coming new high-tech areas you may not have considered. Such futuristic practices within local government are and will continue to be part of our new landscape.

The following is a list of state-of-the-art and innovative technology applications that will require local government standard practices and protocols.

1. Using new and innovative procedures and techniques for **capturing real-time data**, including drone technology, remote sensors, satellite technology and light detection and ranging (LIDAR) data.

2. The new procedures and protocols for **social engagement**, including open and transparent government.

3. The ethical questions of police **practices of using geospatial video technology**, including the use of Amazon's Ring* video doorbell and CCTV. This data is being used by police departments to solve neighborhood crime.

---

* Ring is a home security company recently invested in by Amazon. Ring manufactures a range of home security products that incorporate outdoor motion based cameras and doorbells, such as the Ring Video Doorbell. Wikipedia.

4. The practice of **cyber security** includes securing data and databases in local government.

5. The new business and culture of linking **people, technology, systems** and **software**. This procedure of linking and connecting people to services, resources, amenities and one another will become an ever-more urgent task for local government. Responsibly sharing weather feeds, traffic counts, tweets, crime statistics and Waze* information will prove challenging. Today, some local governments are signing contracts with real-time data suppliers, including Waze.

6. **The protocol and procedures for providing location-based information** to residents. On-demand buses and on-call Uber† taxis and other paratransit operations will become part of the landscape of government.

7. **The practice and protocols of deploying geosensors** for observing, understanding and anticipating events. Sensors read and deliver information about pedestrians, air quality problems, noise abatement and climate change.

8. **The practice of interoperability** enabling different systems, information sources and data types to work together. An example would be computer-aided design (CAD), GIS and building information modeling (BIM).

9. The **ethics and responsibility of activities related to addressing social equity** challenges within a community.

   Other new issues to consider include **technology policies and social media**, the **means of securing public data, procedures and protocols for connecting devices to the system of systems, changes to the infrastructure backbone, new data classifications and** establishing standards for GIS solutions such as dashboards and story maps.

Smart Practices, Activities, Protocols and Procedures

A system of systems relies heavily on smart practices and protocols, as well as data standards. The following is a ground-up list of smart GIS and

---

* Waze is a GPS navigation software app. It works on smartphones and tablet computers that have GPS support. It provides turn-by-turn navigation information and user-submitted travel times and route details, while downloading location-dependent information over a mobile telephone network.–Wikipedia

† Uber is a transportation network company headquartered in San Francisco, California. Uber offers services including peer-to-peer ridesharing, ride service hailing, food delivery, and a bicycle-sharing system. The company has operations in 785 metropolitan areas worldwide.–Wikipedia

geospatial practices. We should compartmentalize these smart practices into the six pillars of sustainability:

- **Pillar One: Smart Governance**
- **Pillar Two: Smart Digital Data and Databases**
- **Pillar Three: Smart Procedures, Workflow and Interoperability**
- **Pillar Four: Smart GIS Software**
- **Pillar Five: Smart Training, Education and Knowledge Transfer**
- **Pillar Six: Smart GIS IT Infrastructure**

Pillar One: Smart Governance

We measure everything – why not governance?

–Mo Ibrahim

Figure 2.2 illustrates recommended smart practices, activities, protocols and procedures related to governance in local government.

1. **Identify a champion and sponsor within the organization**

   The key ingredient for geospatial success is leadership. All geospatially successful local government organizations have key sponsors with dynamic and visionary leadership.

2. **Develop a GIS plan or roadmap**

   A GIS Strategic Implementation Plan provides the roadmap for an organization's successful relationship with geospatial technology.

3. **Update the strategic geospatial roadmap**

   The Strategic Plan should be updated annually. Organizations are organic. Its role, vision and functions constantly evolve. The Strategic Plan must be updated to stay relevant to the organization's vision and the practical aspects of implementation.

4. **Develop an annual detailed GIS work plan**

   A work plan details the schedule and budget for specific projects and initiatives. It not only offers a step-by-step description of the ways a plan will be enacted but projects a timeline and explains how funding will be deployed within the plan's framework.

5. **Develop geospatial GIS vision, goals and objectives**

   The vision, goals and objectives of GIS technology must align with the organizations' vision and also have measurable objectives.

6. **Develop, formalize and (most importantly) ratify a geospatial governance model**

Smart GIS Governance Practices, Activities, Protocols and Procedures
1. Identify a champion and sponsor within the organization
2. Develop a GIS plan or roadmap
3. Update the strategic geospatial roadmap
4. Develop annual detailed GIS work plan
5. Develop geospatial GIS vision, goals, and objectives
6. Develop, formalize, and most importantly ratify a geospatial governance model
7. Establish geospatial job classifications and review job descriptions annually
8. Coordinating enterprise GIS
9. Create and maintain a GIS steering committee
10. Create and maintain GIS technical committee
11. Create and maintain GIS functional groups
12. Create and maintain a GIS user group
13. Develop regionalization strategies for GIS
14. Develop user sensitivity tools
15. Develop GIS collaboration strategy
16. Develop tools to measure quality of GIS service
17. Identify all GIS and geospatial grants and funding initiatives
18. Create GIS based key performance measures or indicators (KPI)
19. Plan, design, and implement a GIS blog or newsletter
20. Promote a culture of GIS collaboration
21. Align GIS with the organization's vision, goals, and objectives
22. Develop service level agreement (SLA)
23. Create a cost recovery strategy
24. Develop a policy regarding revenue generation from GIS products and services

FIGURE 2.2
Smart GIS governance practices, activities, protocols and procedures.

A governance model lays out lines of responsibility and the hierarchy of decision-making within an organization. "Formalizing a governance model allows an organization to maximize accountability and efficiency. It designates the tasks each organizational entity must accomplish" (Holdstock, 2017).

The future of geospatial technology is contingent upon local governments' management of the necessary knowledge, know-how and equipment. I anticipate a future that requires regionally tailored approaches to GIS and geospatial technology.

In the near future, older GIS governance models will become extinct, thereby leaving room for regional variations. Today, there are numerous GIS governance models at work in local government, including:

- Decentralized model
- Centralized model

- Hybrid model
- Hybrid with centralized characteristics
- Hybrid with decentralized characteristics
- Regional model with hybrid characteristics
- Regional model

The Future GIS Governance Model Practice

A hybrid-regionalized GIS Governance Model may well be the model and standard practice of the future. A hybrid-regionalized GIS governance model is based on shared services. In this context, shared services can be defined as "two or more local government authorities that employ staff to oversee geospatial technology including software, management, business and regulatory activities, and IT infrastructure." The essential component in this model is services rendered to communities within the consortium by a multi-agency cooperative agreement. In such an agreement, there is always a lead organization in charge of all geospatial components. This lead organization is charged with "enabling" the participating agencies by planning, setting a direction, hardware, software, training and education. Oftentimes, the lead agency becomes a regional center for GIS excellence. Such multi-agency or regional collaborative activities are organized through a centralized core group of specialists. The ultimate goal of a hybrid-regional GIS model is to have a core nucleus of GIS specialists within the lead organization that enables all other participating agencies. This regional strategy can be ratified in a variety of ways, ranging from simple written agreements (such as an exchange of letters) through loosely structured regional organizations of councils (ROCs) or more formal, legally binding approaches. A regionalized model can often be an extension of a local government's centralized, decentralized or (more likely) hybrid GIS governance model. It essentially incorporates other municipalities and agencies into a shared services model. Examples of multi-agency regional GIS governance models include:

- Pulaski Area GIS (PAGIS), Arkansas
- Muscatine Area Geographic Information Consortium (MAGIC), Iowa
- Clark County Consortium of GIS, Kentucky
- San Diego County Geographic Information Source (SanGIS), California
- Lane Council of Government (LCOG) – Regional Land Information Mapping Database (RLID), Oregon

- Johnson County Automated Information Mapping System (AIMS), Kansas
- King County GIS, Washington
- Gwinnett County GIS Community Partnership, Georgia

Characteristics of a Hybrid-Regionalized
GIS Governance Model

A. Shared costs between participating agencies

Includes the sharing of costs associated with hardware, software, training and education, software development and overall management.

B. A centralized data warehouse

C. The development of a regional center for GIS excellence

D. A regional approach and philosophy to the following:

    a. Economic development

    b. Public safety

    c. Natural resources

E. A collaborative approach to regional GIS activity and grant applications

F. Regional approach to training, education and knowledge transfer

---

Challenges of a Hybrid-Regionalized GIS Governance Model

Future hybrid-regionalized governance models will face many challenges, including but not limited to the following:

- Number and type of GIS users: The number and type of GIS users in a multi-agency model add managerial complexity. More often than not, GIS is underutilized throughout multi-agency initiatives. The sheer number and diversity of user types can present complex managerial problems.

- Software licensing and software applications: Software licensing, plus the number and complexity of software applications within the multi-agency structure can make GIS overwhelming. GIS technology is multifaceted. From a simple public browser to sophisticated predictive modeling algorithms, the technological tasks required to manage desktop, server, internet, intranet, mobile and online cloud solutions present yet another challenge to a hybrid-regional GIS governance.

- Organization-wide agendas, priorities, politics and constraints: The different departmental, organization-wide and regional agendas play significant roles in the deployment of a regional governance model. Budget constraints of a regional model will more often than not determine the level of GIS success.

- Agency-side education, training and knowledge transfer: The mobilizing and training staff across different departments and different organizations can prove challenging.

- The public involvement and engagement factor: The unparalleled ability of GIS software solutions to support open and transparent government in addition to its facilitation of crowdsourcing initiatives and citizen engagement can add another complicated management issue for the regional approach. Each participating agency will have different priorities for sharing and interacting with residents.

- The personality factor can be a problem with multi-agency solutions: Local government organizations employ a wide variety of public safety, engineers, planning, natural resources, economic development and IT professionals. From the technical guru's perspective, GIS implementation is a straightforward process. The management perspective believes that successful GIS implementation is about operational and technical competence. The sociocultural perspective emphasizes the communal dynamics affected by implementation. Managing the variety of perspectives becomes more complicated in a regional governance model.

- Budgeting and funding the hybrid-regional model – Who pays what?: Anyone who was working during the economic downturn of 2008 understands that some of the most precious factors in deploying and maintaining an enterprise GIS are the budget and funding mechanisms. Whether it comes from enterprise funds, operational funds, or capital funds, or multi-agency funds, the funding for a regional GIS model requires constant management and monitoring. Departmental or agency payback models that fund regional GIS are often cumbersome to manage. Monitoring of the budgetary concerns may further politicize the GIS initiative. Additionally, an organization should consider grants and self-education regarding the art of writing grant applications. This is more complicated for the hybrid-regional model.

- Workflow complexity: Standard operating procedures (SOP) are developed to educate staff and simplify complex but repeatable procedures and workflows. SOPs are part of the GIS landscape and require a detailed understanding of business tasks. Uniform standards are hard to enforce across agencies.

- Mobility, technology and architecture: The deployment of a true regional GIS requires understanding and deploying the correct hardware, networking, software and operating systems. Network communications, software architecture and data security must be addressed. In today's world, "office to field" and "field to office" capabilities are paramount to GIS success. Understanding mobile software and the connective potential it brings to GIS technology is critical to maximizing efficiency. This is more complex for a regional approach to GIS.

7. **Establish geospatial job classifications and review job descriptions annually**

   Job classifications enumerate the skill set, financial worth, decision-making power, hierarchical standing and overall responsibilities of a given position within the organization. Job classifications may require adjustment during the GIS implementation process. The increase in technological and problem-solving skills may warrant a re-evaluation of the GIS job classification.

8. **Coordinating enterprise GIS**

   A **coordinated GIS enterprise** refers to a situation where an organization's GIS governance model allows for a GIS coordinator to oversee and coordinate all GIS projects as if they were part of the enterprise.

9. **Create and maintain a GIS steering committee**

   A **GIS steering committee** is a group composed of top-level organizational leaders and GIS specialists. This group often includes all departmental directors of an organization, along with top financial and administrative officers and the GIS coordinator. The steering committee allocates resources for the organization's GIS needs and determines the schedule, priority and policy issues related to implementation. A coherent **GIS steering committee** is crucial for a smooth implementation process, as it allows direct interfacing between executive decision-makers and GIS experts.

10. **Create and maintain GIS technical committee**

    As the name implies, the **GIS technical committee** oversees all of the technical challenges of deploying an enterprise GIS. It sets standards for the ways that GIS data is gathered, managed and shared in an organization. Most of what this committee does is related to systems architecture and IT infrastructure.

11. **Create and maintain GIS functional groups**

    GIS functional groups are specialized teams within an organization responsible for discussing and overseeing key focus areas including public safety, land management, administration and utilities. Functional groups are created usually when the organization is large and complex. These groups essentially divide the task of the GIS steering committee up into management components. They are, by their nature, narrow in focus and require some degree of expertise from their members.

12. **Create and maintain a GIS user group**

    A **GIS user group** is a cohort of stakeholders who share information and compare experiences with GIS technology for the benefit of all members. A GIS user group is managed by the GIS coordinator and meets frequently, often every month or each quarter.

13. **Develop regionalization strategies for GIS**

    Regionalization is a formal agreement between parties or entities to cooperate. In relation to geospatial technologies, regionalization is the sharing of data, resources, applications, training, education and more between disparate groups of GIS users seeking to pool their resources and achieve similar end-goals. Often memorandums of understanding (MOU) are involved in the regionalization of GIS technologies.

14. **Develop user sensitivity tools**

    User sensitivity refers to the capabilities of a particular GIS technology to fluidly respond to a user's request for information. User sensitivity is an important measure of the relative benefits of implementing GIS technology and can be managed by using questionnaires, one-on-one interviews, GIS user group feedback and more.

15. **Develop GIS collaboration strategy**

    GIS collaboration refers to the productive cooperation between individuals and entities facilitated by the implementation of GIS technology. High levels of GIS collaboration let an organization, or organizations, derive maximal benefits from enterprise GIS technologies. It is both a by-product and end-goal of geospatial technology.

16. **Develop tools to measure the quality of GIS service**

    Measuring the quality of service refers to an organization's capacity to gather feedback on data about geospatial technologies. The quality of service can be examined through questionnaires and interviews or metrics related to user interface and objective goals.

17. **Identify all GIS and geospatial grants and funding initiatives**

   A **funding initiative** allows a government organization to diversify funding for GIS. **Grants** are sums of money that are distributed by governmental entities for specific projects. A local government organization should review all opportunities for grant funding to support the GIS initiative. Also, many local government organizations have what are called "enterprise funds" that can be used for an enterprise GIS.

18. Create GIS-based key performance measures or indicators (KPI)

   KPIs are organizationally ratified metrics that gauge whether and how specific goals are met by an organization. These objective, numeric representations of success or failure in GIS enterprise are crucial when comparing the costs and benefits of the GIS initiative.

19. **Plan, design and implement a GIS blog or newsletter**

   A **GIS blog or digital newsletter** is produced by an organization in order to increase communications around a GIS initiative. It provides transparency and accountability by keeping stakeholders and citizens in the loop through easily accessible media.

   Figure 2.3 illustrates the City of Berkeley, California's, information website.

20. **Promote a culture of GIS collaboration**

   A **culture of collaboration** refers to an attitude expressed by stakeholders in their relationships to one another, as they pertain to enterprise GIS. It is an unquantifiable web of positive interpersonal interactions that facilitates creative problem solving and resource sharing amongst individuals and departments to achieve commonly held goals.

21. **Align GIS with the organization's vision, goals and objectives**

   The enterprise GIS needs to be **aligned with the organization's vision, goals and objectives;** otherwise it serves no purpose. This is necessary from the ground up. The vision of an organization may be as simple as improving life for its citizenry. Enterprise GIS supports this vision by identifying areas that need improvement and giving decision-makers the capacity to set realistic, data-backed goals (such as the improved emergency service response time). These goals would then be broken down into objectives to be measured by KPIs.

FIGURE 2.3
City of Berkeley's information website.

22. **Develop service-level agreement (SLA)**

    SLAs are formal, legally binding agreements that outline what stakeholders can expect from enterprise GIS. The parameters of an **SLA** are defined by the KPIs relevant to the technologies in question. Essentially, SLAs can be created to document how the GIS group will support each department.

23. **Create a cost recovery strategy**

    Cost recovery is exactly as the name implies. A cost recovery policy within an organization mandates that the organization will recover the costs for the act of responding to citizen and business requests for data. It may include staff and computer time, as well as hardware expenses including thumb drives or CDs. Essentially, the organization is not charging for data – it is recovering the cost of making it available.

24. **Develop a policy regarding revenue generation from GIS products and services**

    Revenue generation is a policy whereby an organization can actually charge for GIS data and services. Essentially, this is when an organization charges a fee beyond simple cost recovery.

The philosophy is that the set for GIS services can essentially pay for the entire cost of implementing and maintaining a GIS program.

Pillar Two: Smart Digital Data and Databases

We're entering a new world in which data may be more important than software.

– Tim O'Reilly

Figure 2.4 illustrates recommended smart practices, activities, protocols and procedures related to data and databases in local government.

Digital data management and life cycle practices have never been more important. The trend in the geospatial world is a shift toward a centralized geospatial hub of data, decisions, engagement, analysis, visualization and dissemination. This new evolving strategy will rely heavily on the following smart data practices:

1. **Data acquisition**

    Smart real-time and innovative ways of **capturing** vast amounts of digital data is part of our local government landscape today. This is supported by many classic and traditional data capture practices and principles.

Smart Digital Data and Database Practices, Activities, Protocols, and Procedures	
1. Data acquisition	11. Maintain and publish maintenance schedule for critical data layers: a detail maintenance workflow
2. Data and databases standards	12. Detail maintenance workflow and updates for department specific layers
3. Data quality and value strategy	13. Develop enterprise database design standards using Esri's Local Government Information Model (LGIM) and other state and federal guidelines
4. Data analysis, big data analytics and geo-enabled processes	14. Detail all data creation procedures and protocols
5. Data distribution, dissemination and propagation	15. Plan, design and implement a central geospatial repository
6. Data maintenance and upkeep	16. Detail all data stewards or custodians
7. Data visualization	17. Develop mobile solutions standards
8. Regularly conduct a digital data assessment and review	18. Develop a strategy for Arc GIS Hub and open data/ open government
9. Create a master data list and distribute throughout the organizations	19. Data visualization Procedures
10. Plan, design, and enforce metadata standards	

FIGURE 2.4
Smart digital data and database practices, activities, protocols and procedures.

2. **Data and databases standards**

   Data standards, data format, definition and **structure** continue to be an important part of a local government GIS initiative. Standardized digital data models are extremely important to local government. Esri's Local Government Information Model (LGIM) and the Next Generation 911 (NG911) standards as well as Federal Damage Assessment data models are important for local government success.

3. **Data quality and value strategy**

   The **accuracy, validity, reliability, timeliness, relevance** and **completeness of data** are becoming ever more relevant. Using new technology and tools to collect real-time data will become a major part of the duties of a GIO. The quality of geospatial data is critical to making informed decisions.

4. **Data analysis, big data analytics and geo-enabled processes**

   Understanding and analyzing the *when, where* and *what*, as well as the act of **predicting** itself is on an upward trajectory. Dashboard statistical reporting practices require new ways of analyzing and presenting geospatial data with tools like bar charts, line diagrams, pie charts, histograms, scatter plots, dot plots, time series graphs and trend analysis.

5. **Data distribution, dissemination and propagation**

   The hallmark of a true enterprise, scalable and sustainable GIS in local government is the successful and responsible **distribution and dissemination** of data using smart desktop, web, mobile, hosted, virtualization tools, cloud technology and data mashup practices.

6. **Data maintenance and upkeep**

   The practices and protocols of **managing, monitoring, updating** and **maintaining** digital data layers is a vital undertaking and responsibility for local government.

7. **Data visualization**

   The move toward **data visualization** has been relatively new. Local government has been introduced to new tools that offer data techniques to present data in meaningful ways. Dashboards and story maps and analytics tools to find patterns within data are becoming increasingly important.

8. **Regularly conduct a digital data assessment and review**

   A **digital data assessment** examines an organization's existing digital data layers. It evaluates the accuracy, completeness and overall health of the existing digital data layers within an organization.

9. **Create a master data list and distribute throughout the organizations**

   The **master data list (MDL)** enumerates all of the data sets an organization needs for enterprise GIS implementation. The various data sets should be detailed by type and source, and assessed in light of their quantities, accessibility and formats.

10. **Plan, design and enforce metadata standards**

    Metadata describes the collective characteristics of aggregated data. In short, **metadata** is "data about data" (National Information Standards Organization) and distills the patterns evidenced by large quantities of data into more manageable data sets.

11. **Maintain and publish maintenance schedule for critical data layers: A detailed maintenance workflow**

    In the context of geospatial technology, **a data layer** is the visual expression of accumulated data of a particular type. Elevation, city limits or railway lines are all examples of data layers. **Critical data layers** refer to the data layers central to the GIS initiative.

    - **Parcels standards and maintenance workflows:** A **parcel** is a legally defined area of land. A legal description of parcels of land for tax purposes usually accompanies a GIS parcel layer.
    - **Address points standards and maintenance workflows:** An **address point** is a location marked by its position relative to a roadway. An address point is not necessarily the same as a street address. It is a data-point assigned to a mapped location according to parameters that may or may not coincide with a street address.
    - **Street centerline standards and maintenance workflows:** The **street centerline** is a linear data layer that correlates to a center of the roadway.
    - **Aerial photography standards and update workflows: Aerial photography** describes birds-eye-view style photographic data gathered from a plane-, drone- or helicopter-mounted camera. Because **aerial photography** produces an actual image of the mapped terrain, it improves the comprehensibility of practical details.

12. **Detail maintenance workflow and updates for department-specific layers**

    Department-specific layers are mapped representations of data that correlate to the goals and objectives of a single department. For example, the position of every firehouse in a municipality would be **departmentally specific** to the city's Emergency Response Services. Oftentimes, there are hundreds of digital data layers within an organization. Each layer can be specific to a department.

13. **Develop enterprise database design standards using Esri's Local Government Information Model (LGIM) and other state and federal guidelines**

    Enterprise database design refers to the way an organization crafts its data repository in order to meet objectives and further organizational goals. Enterprise database design usually includes focusing on the data, employing data models (Esri's Local Government Information Model (LGIM)) and integration strategies. The design specifies how an organization will collect, share and act upon the various data to produce the desired information products.

14. **Detail all data creation procedures and protocols**

    Data creation procedures are the standardizing guidelines by which an organization's data is collected, catalogued and turned into information products. This is an important set of procedures, as it protects against redundancy and needless work, both of which reduce overall cost-effectivity.

15. **Plan, design and implement a central geospatial repository**

    A **central repository** is an organization's aggregated collection of new and existing GIS data, gathered from all information resources. Pooling data in this manner allows for ease of maintenance, monitoring and collection of metadata. A central repository of GIS data is a characteristic of an enterprise solution.

16. **Detail all data stewards or custodians**

    A **data steward** is responsible for administration and upkeep of specific digital data layers. They are custodians in that they monitor the accuracy and security of departmental data.

17. **Develop mobile solutions standards**

    Mobile solutions refer to GIS applications made available to users via mobile device. In this day and age, mobile solutions are generally geared toward tablet and smartphone users. Figure 2.5 is an example of the City of Berkeley's mobile strategy.

18. **Develop a strategy for Arc GIS Hub and open data/open government**

    Open data and open government describes an increasingly prevalent policy that allows citizens, stakeholders, and non-stakeholders access to an organization's GIS-based data and data layers. Taxpaying citizens can see the results of a GIS initiative. Thus, a more transparent and **open government** is the end goal of this policy.

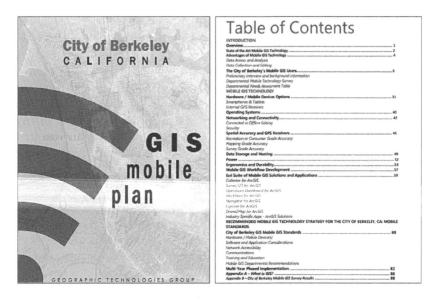

FIGURE 2.5
City of Berkeley, CA: Smart mobile GIS strategy.

19. **Data visualization procedures**

    Demand is rising for GIS software tools that allow users to visualize data and cartography. It is critical to develop procedures and standards for visualizing data in bar charts, pie charts, line graphs and more.

Pillar Three: Smart Procedures, Workflow and Interoperability

The sculptor produces the beautiful statue by chipping away such parts of the marble block as are not needed – it is a process of elimination.

– Elbert Hubbard

Figure 2.6 illustrates recommended smart practices, activities, protocols and procedures related to procedures, workflow and interoperability in local government.

Interoperability will be a defining factor in the future of local government. Interoperability is gradual and phased. It will be influenced by the political, organizational, logistical, tactical, human and technical components of local government.

1. **Develop a strategy for enterprise integration and interoperability**

    Enterprise integration describes the process whereby smaller disparate systems are integrated into the corporate initiative.

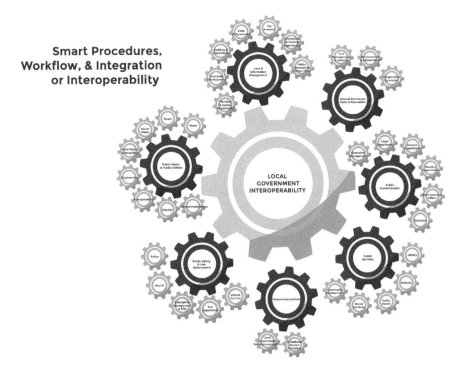

Smart Procedures, Workflow, & Integration or Interoperability

FIGURE 2.6
Smart procedures, workflow and interoperability practices, activities, protocols and procedures.

In a geospatial context, enterprise integration not only encompasses the ways that information moves from departmental systems to the central data repository but also the ways that the new and larger system scope will alter stakeholder relationships and responsibilities.

2. **Detail departmental access to critical data layers**

   Departmental access refers to the ease with which various organizational departments may access these layers. Departmental accessibility is a critical aspect of success.

3. **Create GIS standard operating procedures for data maintenance**

   Standard operating procedures (SOP) are an organization's formally ratified blueprints for actions to be taken in pursuit of a desired objective. They are step-by-step, formulaic and repeatable. In the geospatial context, **SOPs** prevent redundancy in data compilation and unnecessary effort. Adoption of **SOPs** also decreases organizational liability. **Data maintenance procedures** are a subset of **SOPs** that designate how to monitor and keep current the massive amounts of data collected in an enterprise GIS.

4. **Develop GIS "application acquisition/development" procedures**

   A **GIS application** simply refers to the deployment of GIS technologies to generate an information product. **GIS application acquisition/development** procedures are a subset of **SOPs** detailing the ways in which GIS technologies are to be manipulated in order to meet user needs. How does a local government manage software acquisition and/or custom software development? Do these procedures exist? Are they well documented? Does the organization understand the pros and cons of software development?

5. **Define metadata standards**

   It is critically important to **define metadata standards**. Metadata raises political as well as practical issues for enterprise GIS. Clear lines of accountability and quality control for the gathering, storage and application of metadata should be ratified by an organization.

6. **Develop strategy to remove data duplication between systems**

   Data duplication is the actual duplication of data layers. In local government, the most commonly duplicated GIS data layers are street centerlines, address points, parcels and to a lesser extent other boundary layers. Some data layers exist in the databases of three separate departments. However, with the implementation of enterprise GIS, those three data layers are reduced to a single data layer within the central GIS database.

7. **Detail the level of integration and interoperability within the organization – remember system of systems**

   The **level of integration and interoperability** measures how easily technological systems can share, interpret and present data. An effective and enterprise GIS should integrate all databases and offer extensive interoperability. Interoperability means the ability of the GIS to work with other systems within and across organizational boundaries. This includes local, state and federal data sources. The following is a list of key local government enterprise software solutions that require GIS integration:

   - Work order solutions: As the name indicate **work order solutions** manage, process and maintain data about work orders and work performed. Work order solution embrace asset management and GIS centric solutions.

     Figure 2.7 illustrates the importance of understanding the components of Yukon Energy's smart workflow, integration and procedures.

   - ERP solutions (permitting): **Enterprise resource planning (ERP) solutions** are integrative software applications that automate various functions related to planning, permitting, finance and administration.

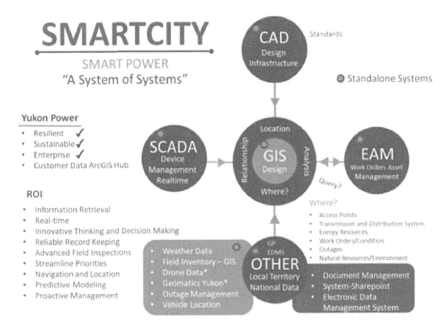

FIGURE 2.7
Yukon Energy: Smart workflow, integration, procedures.

- Public safety solutions: **A public safety solution** is the software application used in computer-aided dispatch, record management system (RMS) and other database and analysis tools.

8. **Develop a strategy for enterprise rather than departmental silos**

   Departmental silos are databases exclusively maintained by a single department. They are full of information and – like actual silos – vertically orientated yet spread out over the terrain of an organization. For example, in a situation with departmental silos, the department of public safety may be the only department that keeps data on crime statistics. In an **enterprise** situation, however, all organizational departments have access to crime statistics via the central database that integrates all departmental data into a single master database.

9. **Develop a GIS technical support (ticketing/help desk) solution**

   Like users of any information technology, GIS users often need help or encounter problems while navigating GIS technologies. The team responsible for an organization's **GIS technical support** will walk users through issues and provide readily available troubleshooting information. Figure 2.8 illustrates an example of a precedence guideline for the City of Boulder Open Space and Mountain Parks organization.

Precedence Guidelines

FLASH	This precedence is reserved for emergent projects of extreme urgency. FLASH contacts are to be handled as quickly as possible, ahead of all other projects. Projects of lower precedence are interrupted until FLASH project is completed
IMMEDIATE	This precedence is reserved for emergent projects that must be completed within 48-72 hours of receipt. IMMEDIATE level emergent projects are handled in order of receipt ahead of all lower precedence level emergent projects
PRIORITY	This precedence is reserved for emergent projects that are to be completed within one work week from receipt. PRIORITY level projects are handled in order of receipt ahead of all ROUTINE level projects
ROUTINE	This precedence is reserved for on-going services and emergent projects with a deadline more than one week out. ROUTINE level projects are handled in order of receipt
PLANNED	This precedence is for standard projects that are long-term, and scheduled in Compass. (ex. Implementation of Beehive Software.)

FIGURE 2.8
OSMP decision matrix.

10. **Document and detail departmental use of GIS**

    This is the actual utilization of GIS within all departments of local government. In the context of geospatial technology, **departmental use** implies a de-centralized implementation of GIS technologies. This model warrants examination of how effectively departments are deploying the technology for different ends. Figure 2.9 graphically illustrates the number of GIS users within the City of Hobart.

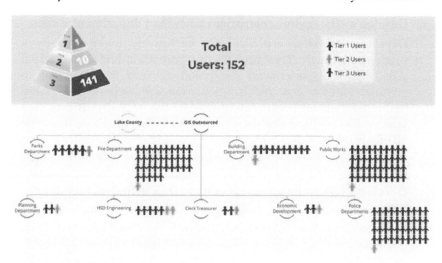

FIGURE 2.9
City of Hobart GIS use.

Pillar Four: Smart GIS Software

We shifted our philosophy from being a computer mapping group that would support planners to the idea of building actual software that would be well engineered. Because at that time, our software was not well-engineered at all; it was basically built with project funding and for project work, largely by ourselves.

– Jack Dangermond

Figure 2.10 illustrates recommended smart practices, activities, protocols and procedures related to GIS Software in local government.

GIS software refers to the network of programs and applications housed on mainframes, servers and the cloud that are deployed to analyze, present and draw conclusions from geospatial data. The end-user interfaces with GIS technology via this software.

1. **Optimized GIS software license agreement**

   A **license agreement** is a legal agreement entered into by the organization and a GIS software vendor that stipulates the limitations, liabilities and appropriate applications of the vendor's technology. An **enterprise agreement (EA)** permits deployment of a software program that is both organization-wide and ceiling-less in terms of

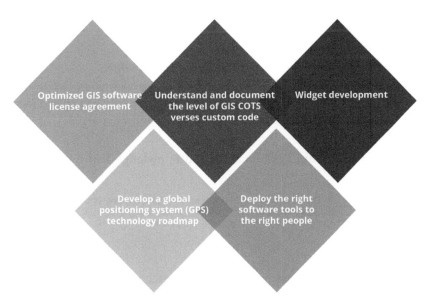

FIGURE 2.10
Smart GIS software practices, activities, protocols and procedures.

user, data or hardware restrictions. The objective here is to measure how available and pervasive GIS software is throughout the organization and create an optimized and cost-effective licensing strategy.

2. **Understand and document the level of GIS commercial off-the-shelf (COTS) versus custom code**

   GIS COTS is GIS software that is commercial-off-the-shelf software. Esri is the de facto local government standard and offers a comprehensive tool set for towns, cities and counties. The objective here is to evaluate how effectively a local government organization uses COTS in developing custom GIS code. Open source code is, as the name implies, code that can be used to create applications for local government. This would be a custom code strategy. However, GIS consultants use custom code to develop solutions that are essentially COTS.

3. **Widget development:**

   Widget is a term for a small software program that augments the functionality of a larger software program. **GIS widgets** provide a way to customize applications in accordance with the specific needs and circumstances of an organization.

4. **Develop a global positioning system (GPS) technology roadmap**

   Global positioning system (GPS) technology is a navigational system enabled by a network of satellites orbiting the earth. The satellites are constantly broadcasting their positions in the sky so that a GPS receiver on Earth can pick up these signals and self-triangulate according to the information received. Though people often get **GIS** and **GPS** confused, **GPS** is a single, though important, tool on the belt of GIS technology. GPS can be used for gathering and monitoring geospatial data.

5. **Deploy the right software tools to the right people**

   One of the most difficult practices in local government is to design the GIS software ecosystem. In today's climate there is an abundance of desktop, web and mobile solutions that can be deployed.

Pillar Five: Smart Training, Education and Knowledge Transfer

The more that you read, the more things you will know, the more that you learn, the more places you'll go.

– Dr. Seuss

Figure 2.11 illustrates recommended smart practices, activities, protocols and procedures related to GIS training, education and knowledge transfer in local government.

| Elearning | Instructor Led In-Person & Viral | Self Study & On-The-Job | Video & Animation | Performance Support |

FIGURE 2.11
Smart GIS training, education and knowledge transfer.

GIS training is the action of *teaching a particular* skill or new type of behavior. Training tends to be more formal and often includes computer technology. GIS education is the *enlightened experience* that follows systematic instruction and usually occurs in an academic setting. Education is less formal than GIS training and does not include anything but the student's presence.

1. **Formal on-going GIS training plan**

   A **formal on-going GIS training plan** is a ratified outline of steps, schedules and costs for continuing to train an organization's employees. It is important to have an **ongoing training plan**, considering that GIS is a rapidly evolving technology and organizational needs are in constant flux.

2. **Multi-tiered GIS software training**

   Multi-tiered GIS software training refers to a standardized process for training employees in the use of GIS technology. **Multi-tiered training** is defined by four distinct types of GIS users.

3. **Mobile software training**

   Mobile software training is the process of teaching users how to engage with GIS technology on their mobile device.

4. **Departmental specific education**

   Departmental specific education provides specialized training procedures according to a department's specific needs.

5. **Return on investment (ROI) workshops**

   Specific workshops related to the value and **ROI** that GIS offers to each department is an important component in the success of an enterprise GIS.

   Figure 2.12 illustrates a flyer for a local government ROI workshop. We can anticipate more of this type of education.

6. **Knowledge transfer**

   Knowledge transfer refers to the process of communicating GIS know-how among different entities in an organization. It is the art

FIGURE 2.12
Flyer for local government ROI workshop.

of *transmitting learned knowledge* from one part of the organization to another. Knowledge transfer is usually accomplished in a very relaxed atmosphere.

7. **Conferences**

   Conferences are gatherings of employees that provide a smorgasbord of opportunities for GIS education. Talks, lectures, lessons and socialization with other industry professionals are ways to advance an understanding of geospatial technologies and keep abreast of new developments.

8. **Online seminars and workshops**

   Online seminars and workshops are online programs implemented by a variety of organizations that further GIS education amongst employees.

9. **Brown bag lunches**

   Brown bag lunches are as informal as they sound. This term refers to a free-and-easy meeting, generally held over a meal where employees can discuss concerns with GIS in a social setting.

10. **GIS succession planning**

    Succession planning refers to an organization's strategy for filling essential but vacant positions with experienced employees. **Succession planning** is a process for identifying and developing new professionals who can replace old professionals when they leave, retire or die. **Succession planning** increases the availability of experienced and capable employees that are prepared to assume these roles as they become available. Succession planning in the GIS world is more difficult than in many other fields.

11. **Cartographic standards**

    Cartography is as much about style and artistic design as it is about information and communication. While there are very few hard-and-fast rules for cartography, there are many standards or conventions that are widely adopted. It is important to take advantage of these commonly recognizable standards in order to make maps readable for end-users.

12. **Quality assurance and quality control**

    Quality assurance (QA) is a set of activities intended to establish confidence that quality requirements will be met. QA is one part of quality management. **Quality control (QC)** is a set of activities intended to ensure that quality requirements are actually being met. QC is one part of quality management.

13. **GIS data disclaimers: General, online and parcel disclaimers**

    Digital and hardcopy maps are created from subsets of GIS databases. Local government often make no claims, no representations, and no warranties, express or implied, concerning the validity and the reliability or the accuracy of the GIS data and GIS data products. A **data disclaimer** is a standard practice and part of the normal procedure of local government. The following language is important and can include: *this map product makes no warranty, representation or guaranty as to the content, sequence, accuracy, timeliness or completeness of any of the database information provided, or the town, city or county is not be liable for any damages, including loss of data, lost profits, business interruption, loss of business information or other pecuniary loss that might arise from the use of this mapping service.*

Pillar Six: Smart GIS IT Infrastructure

Technology is best when it brings people together.
                              – Matt Mullenweg, Social Media Entrepreneur

Figure 2.13 illustrates recommended smart practices, activities, protocols and procedures related to IT Infrastructure in local government.

Infrastructure refers to the network of structures, both physical and systemic, that support an organization's activity. The following is a list of key

Smart GIS IT Infrastructure	
1. Strategic technology plan	7. Enterprise back-up
2. GIS architectural design	8. Data storage
3. Information technology (IT) infrastructure	9. IT, hardware, and mobile standard
4. IT replacement plan	10. GIS mobile action plan
5. GIS training for IT professionals	11. GIS staging and development zone
6. 24/7 availability	

FIGURE 2.13
Smart IT infrastructure practices, activities, protocols and procedures.

GIS infrastructure action items that underpin enterprise, sustainable and enduring GIS solutions in local government.

1. **Strategic technology plan**

   A **strategic technology plan** describes an organization's current and future relationship with technology, and outlines how this technology will further the goals of the organization.

2. **GIS architectural design**

   GIS architectural design is the plan that addresses GIS software technology, capacity performance and IT infrastructure including hardware, network communications, software architecture, enterprise security, backup, platform performance and data administration.

3. **Information technology (IT) infrastructure**

   IT infrastructure refers to a dynamic web of processes, networks, hardware and software resources that support the activities of an integrated IT department.

4. **IT replacement plan**

   An **IT replacement plan** is a formal plan for updating hardware and software resources in the future. Budgetary concerns, goals and long-term objectives are taken into account.

5. **GIS training for IT professionals**

   In order for **IT professionals** to assist an organization with **crowd-sourcing** or **tech support**, they need proficiency in GIS technologies.

6. **24/7 availability**

   24/7 availability refers to the availability of IT infrastructure and GIS technology available at all hours of the day, every day of the week.

7. **Enterprise backup**

   Enterprise backups are a protective measure that preserve an organization's centralized data via offsite, cloud-based daily backup procedures.

8. **Data storage**

   Data storage refers to the digital information storage locally and on the enterprise network.

9. **IT, hardware and mobile standard**

   IT, hardware and mobile standards refer to the formalized set of guidelines and requirements required by the organization to support an enterprise GIS.

10. **GIS mobile action plan**

    A **mobile action plan** is an outline of the tactics an organization will deploy in order to increase GIS accessibility on tablets and smartphones.

11. **GIS staging and development zone**

    A **development zone** is an online site where newly developed GIS applications are tested and tweaked. A **staging zone** is a site where GIS applications are given full trial runs.

History will view the years leading up to 2020 as a period of creativity and innovation. Smart geospatial technology standard practices, activities, procedures and protocols are incredibly important to the present success of local government. Defining and re-defining smart cities, smart government and the geosmart organization is our challenge for the future. Lest we forget, the most fundamental aspect of this progression is the science of creativity and challenges of cultural change. In the next decade, we will have to answer the question: what is the city of our future?

3

A Geospatial Technology Ecosystem: A System of Systems

The new source of power is not money in the hands of a few, but information in the hands of many

– John Naisbitt

Local Government: Towns, Cities and Counties

The world of local government is becoming ever more complex and multi-faceted. There is more data collected, more questions to answer and more options to consider.

More often than not, the vast amounts of digital data collected by local government contain a location or "where" component. Virtually overnight, the location component became both accessible in real time and a vital element of the decision-making process.

All areas of local government value the question: "where?" After all, William Huxhold's 1991 book *An Introduction to Urban Geographic Information Systems* published by the Municipality of Burnaby, British Columbia, reported that eighty to ninety percent of all the information collected and used was related to geography (Huxhold, 1991). Today, local government is a system of elected officials that represent a community. These officials make important decisions about the effective and efficient provision of services to meet community needs. They regulate government activities to pursue the vision, goals and objectives of the community. Local government officials and employees are responsible for providing many important services to the community, including but not limited to:

- Improving working and living conditions
- Supporting the local economy
- Improving the delivery of services
- Promoting a healthy social and cultural life

- Maintaining safety and security
- Improving the level of awareness of citizens about their community
- Maintaining and protecting public property
- Protecting and improving local physical surroundings

The term "local government" in the United States refers to governmental jurisdictions below the state level. These include villages, towns, cities and counties. Local government is made up of divisions and departments that specialize in areas of community services and engagement, including:

- **Public Safety and Law Enforcement**: Police, Sheriff, Emergency Management and EOC, Fire Department, Animal Control
- **Public Works and Public Utilities**: Water, Sewer, Storm Water, Solid Waste and Recycling, Engineering, Transportation, Electric, Telecommunications
- **Land and Information Management**: Planning and Zoning Department, Economic Development, Building and Inspections, Code Enforcement, Tax Assessor, Information Technology Department, Public Information Officer (PIO)
- **Natural Resources, Parks and Recreation**: Tree Management and Arborist, Environmental and Conservation, Cooperative Extension
- **Public Administration**: Executive Management, Legal Department, Finance Department, Housing Department, Environmental Affairs, Elections
- **Public Services**: Library, Schools, Public Health, Social Services, Community Development
- **Telecommunications**: Local Government Telecommunications, Broadband Service Providers

Today's local government organizations operate at different levels of economic and technological status. Philosophies about information technology vary from municipality-to-municipality and state-to-state. In spite of these factors, the gradual and noticeable shift toward geo-smart government and the city of the future has already begun.

So what are the main components of a city of the future?

The Shifting Challenges of the City of the Future

The city of the future will face specific urban challenges including but not limited to the following:

Challenge #1: The Ecology of a City: Natural features and forces play crucial parts in the urban landscape. Protecting wildlife, habitats and natural resources will remain important part of good stewardship. In the future, there will be a major emphasis on reducing impacts on the natural ecosystem. This emphasis will be supported by sustainable technologies and processes that include green roofs, roof gardens, absorbent rain gardens, solar panels and small-scale farming within the future city. Cities that have already embraced this mindset include:

- City of Boulder, Open Space and Mountain Parks (OSMP), Colorado
- City of Edina, Parks and Recreation Department, Minnesota
- Hoffman Estates Park District, Illinois
- City of Berkeley, California, Parks and Recreation Department

Challenge #2: Preservation and Protection of Water Resources: The protection of water systems, the collection and cleansing of storm water and improving overall water quality are three of the future city's most important tasks. Wetland restoration, flooding and sea-level rise also pose serious issues for our future. We will see an embrace of rainwater cleansing, plus smart water and sensor technology that maximizes irrigation efficiency in city farms. Cities that have already embraced these measures include:

- Contra Costa Water District (CCWD), California

Challenge #3: Renewable and Clean Energy: Renewable energy is a hot topic. The urban building of the future will have to generate as much energy as it consumes. The production of energy will function in close proximity to the end consumer. Cities that have already recognized this necessity include:

- Yukon Energy Yukon Territory, Canada
- City of Lawrenceville, Georgia
- City of Healdsburg, California

Challenge #4: Waste and Recycling: We should anticipate the near-future arrival of fully automated waste collection, recycling and reuse. Waste is a resource that can produce energy and alternative materials. Automated recycling will become part of our everyday activity. Smart waste technologies will play a significant role in our future city landscape.

Challenge #5: Food and Urban Farming Solutions: We will see significant increases in urban farming and hydroponic technologies. Sustainable food generation practices will be maintained across the life cycle of a product. Improvements in food production, food delivery and disposal as well as organic farming standards underpin this aspect of our future. Locally sourced food will continue gaining momentum and integrating itself into our culture. One of the largest benefits of urban agriculture is the reduced shipping distance between farmer and buyer. Container farms will become a part of the city landscape, where old metal shipping containers from cargo ships are converted into mobile farms.

Challenge #6: The Mobility of People: Future mobility in and around our cities will hinge upon safe and affordable high-speed rail. Urban areas will have more pedestrian space. Cities will become walkable communities. There will be fewer personal vehicles within cities due to improved vehicle monitoring and traffic congestion analysis. Autonomous vehicles, too, will impact road infrastructure and safety. Our physical landscape will change as signs, lights and other manual vehicle infrastructure are removed. Government organizations that have already embraced this mindset include:

- The Country of Qatar
- Ashghal Public Works Authority, Qatar

Challenge #7: Protecting Our Culture and Heritage: The celebration of history and heritage will only become more important. Recreation, arts, community and cultural diversity will be significant elements of the city of the future. The historical and cultural heritage of cities will be promoted and maintained by means of smart city tools, virtual and augmented reality and various applications.

Cities working with this type of mindset include:

- Dutch Harbor and Native Ounalasshka corporation, Alaska
- Seminole Tribe Of Florida (STOF)

Challenge #8: The Livability of Our Cities: The future city will require safe and streamlined access to nature, services and automated technology. This will require strategies that collate disparate livability factors such as amenities, demographics, economy, education, health care, housing, social capital, transportation and infrastructure. The LivScore methodology (a strategy to assign a livability value to a community) is an example of how a composite of more than forty data points can be grouped into the eight categories to assess the livability of a city. Cities already working with this type of mindset include:

- City of Lawrenceville, Georgia
- City of Napa, California
- City of Berkeley, California
- City of Mississauga, Ontario, Canada

Challenge #9: Public Safety and Hazard Mitigation, Emergency Planning, Response and Recovery: The city of the future will require innovative public safety strategies. Real-time incident data and decision-support technologies will mitigate natural and human-made hazards in large urban areas. Cities already working within this mindset include:

- Bay County, Florida
- Washington County multi-agency strategies and Carteret County, North Carolina

Challenge #10: City Infrastructure: In the future, urban buildings will be designed for pedestrian use and constructed with natural resources that effectively utilize water, soil and air. Water, sewer, storm water, electric, gas and telecommunications infrastructure will be tracked, measured and monitored using smart sensor technology. Organizations already operating with this type of mindset include:

- The Country of Qatar
- Ashghal Public Works Authority, Qatar
- City of Lawrenceville, Georgia
- City of Dayton, Ohio
- Town of Windsor, California

Challenge # 11: The Economy and Policy: Future economic policies will safeguard economic sustainability. Tomorrow's city will aim to become ever more sustainable and resilient in the face of economic downturns, hazards and demographic changes. Cities that have already embraced this mindset include:

- City of Mississauga, Ontario, Canada
- City of Hobart, Idaho

Challenge #12: Land Use and Zoning: Land use and zoning laws will play significant roles in our urban future. Local government will have to address entirely new requirements about livability of cities, infrastructure, smart building codes and the marriage between industrial, agricultural, commercial and residential. Cities already working with this mindset include:

- City of Irvine, California

Challenge #13: Smart Buildings and Smart Technology:

Buildings will become smarter by using advanced technology and incorporating natural elements into their design. The urban building will serve as residences and yet offer more cross-purpose amenities like gardens and energy efficiency. The future of smart buildings will rely on connected sensors and cloud computing. IoT devices intelligently monitoring and controlling the operations of a building will become common place.

Challenge #14: Self-Contained Resilient and Sustainable Neighborhoods, Regions or Zones: Neighborhoods will be quantified and evaluated in accordance with a sustainability index. This index will measure the effective co-existence of humans and nature and the neighborhood equitability. Cities already working with this mindset include:

- City of Hobart, Indiana – Hobart Sustainable Neighborhood Environmental/Ecological Subplan

Challenge #15: Urbanization, the Urban Hub and the Mega City: By 2050, the world's population will be nearly 10 billion. Seventy percent of humanity will live in urban areas. Managing the shape and form of cities in addition to the urban lifestyle they offer will pose significant challenges. Rail, road, natural boundaries and real estate already play a significant role in the physical shape of cities. The urban form of the future will be structured and monitored using smart technology. The centralized smart data HUB will utilize data and databases from a myriad of sources to support better decision making with regard to planned communities. Cities already working with this mindset include:

- City of Simi Valley, California
- Thousand Oaks, California
- City of Irvine, California
- Part of the Los Angeles urban sprawl of California

Figure 3.1 details the shifting challenges of the city of the future.

The Smart City of the Future

The migration of humans from rural areas to cities is already forcing local governments to embrace innovative new ways of providing sustainable places to live. Cities are – without doubt – slowly but surely becoming technologically smarter.

A smart city is a designation given to a local government organization that is a "connected city." A "connected city" incorporates, integrates and embeds information technology into the decision-support apparatus of local government. In furtherance of its municipal goals, a "connected city" will actively work toward an interoperable "system of systems."

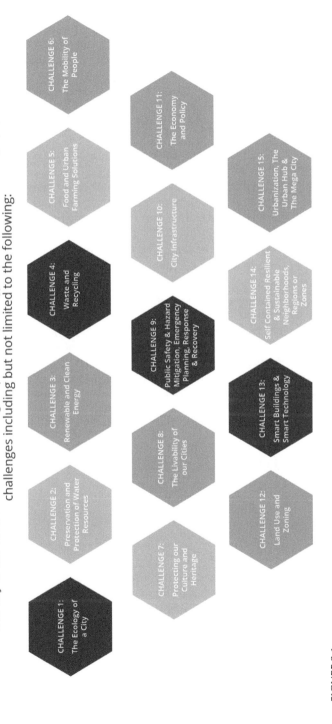

FIGURE 3.1
The shifting challenges of the city of the future.

Smart data is used to enhance, improve and re-engineer operations and the provision of government services. A smart community is a town, city or county that effectively uses and shares electronic data and databases, including the IoT, air quality sensors, stream gauge sensors, parking meter sensors, pedestrian sensors, traffic congestion sensors, drone technology, smart traffic lights, the tools afforded by geospatial technology, open data, smart energy solutions that make residential and commercial building more energy efficient and high-tech data gathering devices that supply information critical for the management of assets, operations, functions, people and resources.

Smart cities are all about big data analytics and predictive analysis. These are new ways of consuming and exploring disparate databases and represent entirely new relationships between data. A smart community is a municipality that focuses on real-time data to increase operational efficiency, share information across an organization and improve the quality of government services.

Additionally, a smart city is a municipality that offers its residents mobile software applications to access and report critical information about traffic, parking, congestion, crime, infrastructure failure, potholes and graffiti. Open data and transparent government is a key element of this model.

Today, it is on trend for local government to develop smart city master plans (SCMP). The SCMP should articulate a smart city mission, vision and goals and objectives. It should include high-level strategies and address sustainable, reliable and resilient solutions. It must be as responsible, relevant and timely as it is equitable, transparent and collaborative. The overall guiding principles require smart choices in the following fields:

- Public safety and security
- Built environment
- Energy and utilities
- Transportation system
- Water and wastewater
- Solid waste management
- People and community

Smart city technology will eventually affect everyone. Europe, Canada and the United States are embracing this concept in order to improve the quality of life in large urban areas. Municipal, state and federal governments – as well as private and public partnerships – are becoming involved in the local government adoption of new smart technology. According to McKinsey research (as published in Tech Republic), "The smart city industry is projected to be a $400 billion market by 2020, with 600 cities worldwide. These cities are expected to generate 60% of the world's GDP by 2025."

The technological environment in local government is broad and varied. The following passage illustrates the current extent of technology in local government. Arguably, the following list of solutions reveals the slow march of local government into the era of smart cities.

Geospatial Science and Technology of the Future

Mapping out our brave new world requires an understanding that our future is always the present. We react and evolve to our existing environment. We do not necessarily predict it. We just evolve at different speeds. Therefore, let's take a look at what we have today in an attempt to anticipate what that solution will be tomorrow and beyond.

Let us start by reviewing what local governments have in terms of Esri-based geographic information systems (GIS) solutions. They include the following:

- Desktop solutions
- Web applications
  - Arc GIS HUB
  - Intranet solutions – Departmental applications
  - Internet solutions – Public facing applications
- Mobile solutions
- Dashboard solutions
- Story maps
- Enterprise architecture tools
- Bespoke GIS initiatives
- Third-party integrations (3rd) and solutions

Integrating the above geospatial solutions appears far simpler than it actually is. There are over fifty products and solutions that are part of this desktop, web, mobile and architectural ecosystem. Conventional wisdom says that GIS and geospatial science evolved in step with all other societal advancements, whether they be technological, social or environmental improvements. For example, the environmental movement of the 1970s was a social development that inspired the GIS world. Equally inspiring is today's smart city concept that fuses both technological and social concerns. Turning our attention to the future of world and local governmental technology may allow us to see far into the future of geospatial problem-solving.

In Michio Kaku's book *The Physics of the Future: How Science Will Shape Human Destiny and Our Daily Lives by the Year 2100*, Kaku talks about the future of computers, artificial intelligence (AI) technology, nanotechnology, energy and humanity itself. Now, ordinarily, I would consider Kaku's book a list of crackpot guesswork. Kaku, however, did something different than a normal prognosticator: he talked at length to all of the leading technology experts. This extensive data gathering allowed him to formulate his ideas and develop a comprehensive and accurate manuscript of our future. Interestingly, Kaku explicitly suggests ranking civilizational success by entropy, otherwise known as measuring energy efficiency. That sounds awfully like sustainability and resilience to my ear.

Kaku and others researchers anticipate that our new advanced technology will include **Internet glasses, contact lenses, driverless and intelligent electric cars, roll up TV and monitor screen for all walls in an office and home, super thin flexible electronic paper,** universal language translators **(ULT), expert systems and robotics, extensive nanotechnology and energy extraction from the sky satellites.**

The next section is a review of technology used in local government today and an explanation of the technology we will see tomorrow.

Today's Local Government Smart Government Technology

Public Safety and Law Enforcement Smart Technology: Police and Sheriff Departments

- **Crime Analysis and Predictive Analysis**: This technology provides law enforcement with visual, real-time crime information securely, time-efficiently and accurately so police can respond before criminals even strike.
- **Smart CCTV Cameras**: This technology involves computer algorithms that continually predict human movement and actions, allowing crimes to be thwarted.
- **Gun & Holster Sensors**: A sensory system is applied to policemen guns and holsters, that registers separate "events" when the gun has been removed, inserted or any part of the gun has been engaged. These events are reported via encrypted Bluetooth to a designated data hub.
- **Body Cameras**: Body cameras help keep all parties accountable, i.e. the police officers and the people they interact with. The cameras archive videos and photos for evidence.
- **Video Doorbell**: Smart doorbells and door viewers detect motion as people approach a property, send notifications and video feeds to the house owner's mobile devices and PC and allow the owner to view and communicate with whoever is at their door.
- **Facial Recognition**: This technology identifies criminals in video footage and photos, then tracks them down so police can thwart crime attempts before they happen.

- **ShotSpotter**: *ShotSpotter* technology relies on acoustic sensors spread throughout an area. When a gunshot is detected, the sensors communicate with the Incident Review Center. The Center then transmits situationally detailed messages to dispatch centers, patrol car Mobile Data Terminals (MDTs) and officer smartphones.

- **License Plate Recognition (LPR) Cameras**: These cameras take photos of license plates. They store GPS information and timestamps in a secure database that can only be accessed by authorized personnel within an agency.

- **Drone Technology**: Drones enhance the surveillance capabilities of police forces. They can assist in the pursuit of suspects and find stolen vehicles. Drones also provide a bird's eye view of the neighborhoods and city streets being patrolled.

Public Safety and Law Enforcement Smart Technology: Emergency Management and Emergency Operations Center (EOC)

- **Real-Time Weather Feed**: Various platforms and weather-forecast companies provide real-time weather data, such as radar and satellite information. Some of these companies provide exclusive data via paid services and support.

- **Reverse 911**: "Reverse 911" is an application that accesses a database of all the telephone numbers of households impacted by a threatening situation and then sends a pre-recorded message to each number.

Public Safety and Law Enforcement Smart Technology: Fire Department

- **Smart Fire and Smoke Detection**: These detectors can send alerts to smartphones. If a resident is not at home, smart fire and smoke detectors will call the fire department. An additional perk: residents can temporarily silence the alarm from a smart phone.

- **Firefighting Drones**: Drones provide fire-ground assessment to help commanders quickly gather critical information on the scene of a fire or incident involving hazardous materials. Drones can aid search and rescue operations and possibly deliver food, water, or first aid kits to places that firefighters can't reach.

- **Personnel Accountability/Tracking Tech**: Advances in this technology allow mobile applications to automatically detect essential information (who is on the scene, their location, their safety status) without requiring manual-active involvement from the commander.

- **Wireless, Environmental Sensors**: A number of wireless technologies for environmental sensors are being integrated into personal protective equipment (PPE) for firefighters:

- **Thermal Imaging Cameras (TIC)**: These cameras pinpoint the source of a fire, help find victims and accelerate the process for finding fires in the void spaces between walls.
- **Gas Dosimeter Systems**: This technology monitors each individual firefighter's exposure to contaminants. It can even narrow down the chemicals to be checked for when decontaminating Personal Protective Equipment (PPE).
- **Heat Flux Measurement Gauges**: These gauges monitor potential flashover conditions by measuring the temperature in the upper layer of air.
- **Personal Alert Safety Systems (PASS)**: Containing an acoustic transmitter (like a beacon), this device detects firefighter motion and locates firefighters when they have been immobile for a predetermined period of time.

Public Safety and Law Enforcement Smart Technology: Animal Control

- **Smart Dog Collars**: These collars are capable of GPS-tracking, Wi-Fi and even cellular data transmission. By sending this information to a smartphone, a smart collar lets the owner know where their dog is and whether their dog is getting enough activity for its health.

Public Works and Public Utilities: Water Department

- **Smart Meters**: These meters measure, store and frequently send data via wireless communication regarding the amount of water consumed in a building and the date and time of consumption. These meters can be read remotely at any time.
- **Smart Water Quality Sensors**: These sensors can be remotely operated and maintained to analyze quality of drinking and fresh water throughout industrial processes.

Public Works and Public Utilities: Sewer

- **Smart Sewer System**: Smart sewers use sensor technology (including radar and sonar technology) to track real-time sewer overflows and optimize the use of its existing sewer system. This type of technology may have the additional ability to detect potential health epidemics, infectious diseases, and predict potential outbreaks.

Public Works and Public Utilities: Storm Water

- **Continuous Monitoring and Adaptive Control (CMAC)**: This technology controls the rate of storm water flow, giving operators the leeway to plan and respond to storm events.

Public Works and Public Utilities: Solid Waste and Recycling

- **RFID Tags**: These tags on a waste or recycling bin can identify and trace waste streams, transmit the number of times a container is placed for collection and track the weight of its contents. Lastly, they simplify the process for billing customers.
- **Wireless Bin Sensors**: These sensors track the fill level of waste and recycling bins, then transmit information over cellular networks to real-time data dashboards.

Public Works and Public Utilities: Engineering

- **Smart Buildings**: These are two intertwining concepts – Building Information Modeling (BIM) and Building Internet of Things (BIoT) – that promote collaboration between the construction sector and the software industry. The designers will integrate artificial intelligence into the building's operation so that it can "learn" from its "experiences" of running the facility and interacting with human facility operators. The United Kingdom is taking a strong lead with regards to BIM and BIoT implementation in the construction of new buildings and the retrofitting of older buildings.

Public Works and Public Utilities: Transportation

- **Smart Street Lights**: These lights incorporate sensors and other technologies for applications related to gunshot detection, environmental monitoring, charging electric vehicles, traffic management and smart parking.
- **Smart Traffic Volume Monitoring**: This technology tracks vehicle volume and traffic conditions by street direction, time of day and day of the week. Additionally, it tracks traffic speeds in different lanes and reports ETAs (Estimated Time of Arrival).
- **Smart Electric Vehicle Charging Stations**: These stations employ cloud-based software installed specifically for the entity using the stations – property owners, residents, municipalities, highway or freeway – and the necessary charging speed.
- **Smart Traffic Management Center**: This type of system integrates traffic signals and sensors to manage the flow of traffic based on pressing situational needs.
- **Smart Parking**: This concept combines technology and human measures to facilitate quicker, easier and denser parking. One certification program, called *Park smart*, promotes sustainable parking facilities, the use of fewer resources and innovative parking-transportation concepts and technologies.

- **Pedestrian Counts**: Sensors are placed in specified locations throughout a city or a park and wirelessly transmit the counts over a network to designated location.

- **Autonomous Cars**: These vehicles are capable of detecting their surroundings and navigating paths and even around obstacles. Notably, they use using radar, LIDAR, and GPS.

- **Dockless Scooters and Bicycles**: These scooters and bicycles connect to mobile phone apps that enable riders to rent vehicles from anywhere. The apps unlock the scooters and bikes for rental then lock up when the renters are finished.

Public Works and Public Utilities: Electric

- **Smart Grid Sensors**: These sensors remotely monitor conditions surrounding electrical equipment, such as power lines and their temperature, relevant weather conditions and the power demand of resources on a smart grid.

- **Smart Electric Meters**: These meters measure and record electric energy consumption hourly (or even more frequently). The meters then remotely communicate the information to the electricity supplier, thus streamlining the billing process.

Public Works and Public Utilities: Alternative Energy Production

- **Trigeneration**: This form of energy production blends cooling, heat and power (CCHP), by taking the heat produced by a typical cogeneration plant and using it to generate chilled water. The benefits of trigeneration include reduced fuel and energy costs, lower electrical usage during the summer and the lack of harmful chemical pollutants produced by a water-refrigerated system.

Public Works and Public Utilities: Fleet Management

- **Smart Fleet Management and Routing**: Companies have a growing need for real-time data availability that allows for reporting traffic conditions, receiving alerts for vehicle problems, the ability to view vehicles enroute in real time and directly communicate re-routing to drivers.

Land and Information Management: Planning and Zoning Department

- **3D Urban Planning**: Well-visualized urban development models will become increasingly necessary for planning and zoning work. There are a number of platforms developing and applications that represent both the real world and architectural construction models using 3D graphics.

Land and Information Management: Urban Innovations

- **Accessing Unused Resources**: In the electric, water, parking, transportation, housing and other industries, many companies keep a sizable amount of resources in reserve in case of peak demand. Decreasing the gap between unused and used assets is essential to mitigate increasing demand for limited resources.

Land and Information Management: Economic Development

- **Marketing & Promotion**: Places such as Memphis, New Orleans, Daytona and San Francisco that have developed names for themselves because of promoting sights to see, local eateries, activities and festivals. Smart cities will follow suit.

Land and Information Management: Building and Inspections

- **Smart Building Sensors**: These sensors are installed on and inside buildings to provide real-time data to facility managers for assessing and tracking infrastructure conditions over time. This will expedite inspections and allow artificial intelligence to integrate with a building's lighting system in order to reduce energy costs.

Land and Information Management: Code Enforcement

- **Local Government-Community Collaboration**: At any one time, there are only so many code-enforcement staffers available for monitoring violations in a given city or town. Enlisting residents to help identify issues with building codes will assist government staff. Companies with mobile applications are already providing the means for residents to send photos of code violations to their local code-enforcement department.

Information Technology Department

- **Enterprise-Class Visibility, Security and Control**: There are methods for analyzing and protecting against threats and securing all technological assets like company-owned smartphones, printers, networks, laptops and other devices. Risks, misplacing of devices and security breaches are some of the greatest threats to important information.

- **Cloud Technologies**: Managing space on devices and servers, securing data, uploading data or working out of the cloud is essential for local governments, departments and organizations.
- **Mobile Internet of Things (IoT) Technologies**: Some wireless providers promote real-time monitoring, control and management of mobile-device use among staff. In doing so, they allow for better connectivity and network communication both domestically and internationally.

Natural Resources, Parks and Recreation: Parks and Recreation

- **Smart Parks**: There are a number of new concepts and ideas being researched to create and refit parks to make them more sustainable, inviting and integrated with smart technology.
  - **Public Wi-Fi**: This is the norm at many fast-food restaurants, coffee shops, airports, and grocery stores. Installing nodes and local Wi-Fi routers for parks is forward thinking.
  - **Smart Lighting**: Smart lighting in parks will be a valuable asset for cutting down on energy costs and reducing energy output. The light posts can have other smart technologies such as security cameras, Wi-Fi routers or environmental sensors installed on them to maximize use of resources without distracting visitors.
  - **Smart Benches**: These are solar-powered benches that have USB and energy outlets for charging visitors' smartphones and other devices. They can also integrate environmental sensors.
  - **Smart Kiosks**: A number of cities are installing these kiosks to display local news updates, tourism information, weather alerts and education information to engage visitors.

Natural Resources, Parks and Recreation: Tree Management and Arborist

- **Smart Monitoring**: Sensors can be placed around natural resources and in the soil to monitor health and growth conditions. These sensors can transmit this data over a private or public network to mobile devices and PCs, thus providing essential information to landscapers and arborists.

Natural Resources, Parks and Recreation: Environmental and Conservation

- **Air Quality Sensors**: There are already innovative products in use that assess the air quality inside businesses or residences based on carbon dioxide levels in addition to temperature and humidity levels, the presence of chemicals or even fine dust. Additionally, smart environmental sensors can be placed outdoors to monitor air quality and transmit this data to devices.

- **River Gauge Sensors**: River gauge sensors monitor discharge, stream-flow, runoff, and other hydrologic features and conditions in areas that are unattended for short and long-term periods. These sensors send data remotely from the site to management centers.
- **Weather Feed & Alerts**: In addition to local National Weather Service Forecast Offices, there are companies providing detailed and consumer-specific alerts and monitoring assets. They may provide 24/7 support for weather-forecast services, emergency preparation and real-time data feeds.
- **Real-Time Weather Stations**: Companies are utilizing weather stations all over the country to monitor temperature, humidity, rainfall, wind direction, wind speeds and other environmental conditions. Some of these companies deploy additional weather stations to provide data to their customers in real time.

Public Administration: Executive Management

- **Employee-Led Training**: Local government management can encourage employees with various areas of expertise to create training programs for fellow employees who are interested in growth and skillset expansions. This promotes opportunities for employees to grow in their leadership and training skills and develops a collaborative environment for the local government staff.
- **Employee Training Academy**: Local government management can survey its employees about how to improve the work environment, culture, life and internal skillsets. Survey results can be used to create a training academy or program.

Public Administration: Legal Department

- **Cloud-Based Solutions**: Store important documents and paperwork in the cloud to protect important information from fire, theft, water damage or loss.

Public Administration: Finance Department

- **Mobile Technology**: Mobile technologies can be used for ordering, processing, and delivering transactions to the local government finance department. This technology reduces lines at the finance and law enforcement centers by allowing residents and businesses to avoid paying in person for services, fees and other things.

Public Administration: Housing Department

- **Housing Frequently Asked Questions (FAQ) Using Artificial Intelligence (AI)**: Rental and housing property owners use virtual assistance to answer tenant FAQs that tenants. Tenants may access this technology at any time. Both parties benefit from reducing the number of calls to the management office.

Public Administration: Elections

- **Electronic Voting**: A number of countries around the world have transitioned from paper ballots to electronic voting methods. They deliver faster results, increased security, fraud protection and a streamlined voting process. Measures can be taken to validate proper voter registration for confirmed citizens.
- **Mobile and Online Voting**: A further step for elections is to enable mobile and online voting. This increases convenience, reduces traffic considerably at ballot centers, and allows for quicker results.

Public Services: Library

- **Makerspace and the Maker Movement**: This movement encourages students, adults and entrepreneurs of all ages to participate Do-It-Yourself and Do-It-With-Others environments and activities. Participants are emboldened to create new things, re-create old things, and improve upon existing ideas. Libraries, schools and other public or private facilities can host Makerspaces.

Public Services: Schools

- **Mobile Incident Reporting to Police**: To increase security, schools can integrate mobile apps for students and faculty to instantly report safety concerns and incidents to local law enforcement and campus police. This allows for rapider deployment and response.

Public Services: Public Health

- **Smart Health**: Easily interconnectable hospital assets allow hospital managers to view a wide variety of data in real time. Smart health innovations improve management of resources, staff, medical devices and enable better tracking of patients' statuses, ultimately allowing for higher-quality care and quicker service.

Public Services: Social Services

- **Block Chain Data Security**: Block chain data blocks hold information in each block. The first block contains basic information (name, date of birth, etc.). Subsequent blocks are added when new prescriptions, appointments, hospitalizations or other medical situations occur. This allows all hospitals, doctor/dental/vision offices and other medical facilities to pull from the same block chain record for each patient.

Telecommunications

- **Array of Things (AoT)**: The concept of AoT is described as connecting urban buildings, neighborhoods, streets, sensors, street lights, and communication networks to provide urban activity information in real time to further research, innovation and development on the transition to smart cities.
- **Transition to Internet Protocol 6 (IPv6)**: According to *The Internet Society*, the number of the current form of Internet Protocol (IP) addresses – IPv4 – will be depleted in 2021. There is a need for operators of organization networks to obtain IPv6 addresses for the purpose of growing their networks. It will become increasingly difficult and expensive to obtain new IPv4 addresses.
- **Smart Communication Networks**: New network communication options exist for specific applications regarding utility use of smart sensors, water and electric meters. These networks ensure privacy and eliminate the potential for radio frequency interference from other users. Advantages include higher transmission power and two-way communication between the utility company and their customers.
- **Smart Fiber Management Solution**: Improved Automated Infrastructure Management (AIM) uses fiber optic enclosures and real-time cross-asset monitoring of an entire network, from office-space areas to server rooms. Real-time monitoring can instantly alert IT and security teams to unsanctioned occurrences on a network.
- **Augmented Reality**: 3D viewers, 3D internet browsers and 3D GPS navigation technologies exist for driving and walking. They are available for smartphones, tablets and other mobile devices.

Real-Time Data Feeds

- **Emergency Management**: Given natural disasters, severe weather and public safety threats, it is crucially important for local governments to access real-time data feeds in order to provide situation

and status updates, safety information and disaster preparation information. The Federal Emergency Management Agency
(FEMA) is a great example of an organization making use of these
data feeds.

- **Business-Organization Analytics**: Dashboards are available
  for analyzing, tracking and monitoring real-time and idle data-
  encompassing assets and financial activities. These dashboards can
  improve supply flow management by simplifying the view of raw
  data in a way that allows for meaningful understanding of important information.

The Geo-Smart City of the Future

We have two things. We have the fifteen challenges the city of the future
will face (as seen in Figure 3.1) and we have a proposed **definition of a
smart government organization of the future**. The next question we need
to answer is: where does geo-smart government fit in to this? How should
we adapt our geospatial thinking to a changing world? Figure 3.2 is a
bold statement about the future of GIS and geospatial technology in local
government.

We should use the six pillars of GIS sustainability in our attempts to
describe and discuss the future of the geospatial industry and geo-smart
governments. Fragmenting the question "what is the future role of geo-
smart government in our future" into the six pillars of GIS sustainability
offers concision and clarity.

The six pillars of GIS sustainability regarding geo-smart government
include:

- Pillar 1: Smart Geospatial Governance Components
- Pillar 2: Smart Digital Data and Databases Components
- Pillar 3: Smart Procedures, Workflow and Interoperability Components
- Pillar 4: Smart GIS Software Components
- Pillar 5: Smart Training, Education and Knowledge Transfer Components
- Pillar 6: Smart GIS IT Infrastructure Components (Please note: The
  term *information technology* (IT) infrastructure refers to the entire
  collection of *hardware, software, networks, data centers,* facilities,
  and related equipment used to develop, test, operate, monitor, manage and/or support geospatial information technology
  services.)

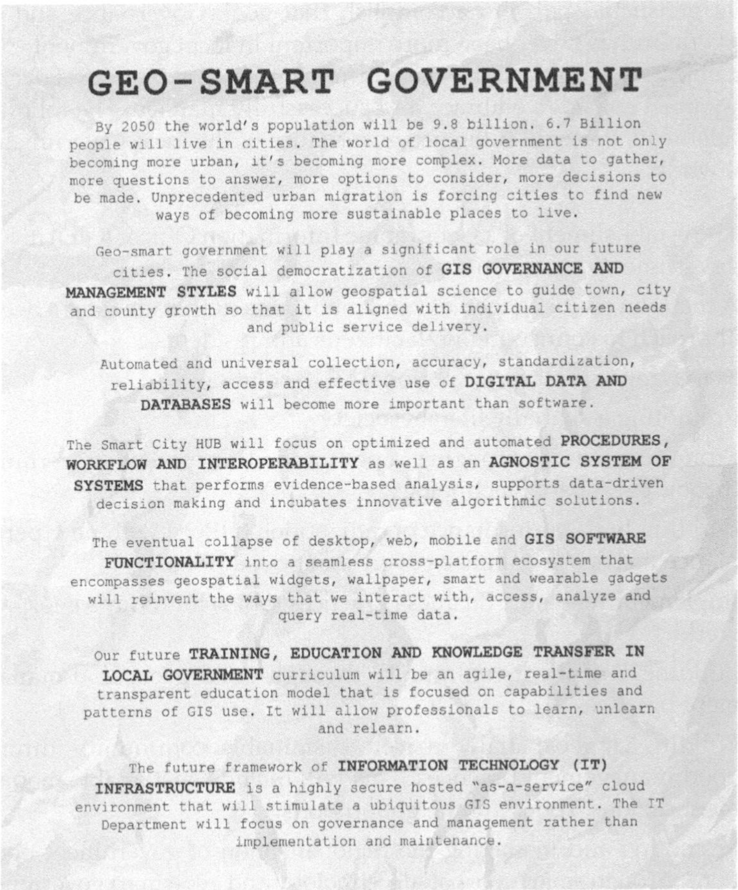

GEO-SMART GOVERNMENT

By 2050 the world's population will be 9.8 billion. 6.7 Billion people will live in cities. The world of local government is not only becoming more urban, it's becoming more complex. More data to gather, more questions to answer, more options to consider, more decisions to be made. Unprecedented urban migration is forcing cities to find new ways of becoming more sustainable places to live.

Geo-smart government will play a significant role in our future cities. The social democratization of **GIS GOVERNANCE AND MANAGEMENT STYLES** will allow geospatial science to guide town, city and county growth so that it is aligned with individual citizen needs and public service delivery.

Automated and universal collection, accuracy, standardization, reliability, access and effective use of **DIGITAL DATA AND DATABASES** will become more important than software.

The Smart City HUB will focus on optimized and automated **PROCEDURES, WORKFLOW AND INTEROPERABILITY** as well as an **AGNOSTIC SYSTEM OF SYSTEMS** that performs evidence-based analysis, supports data-driven decision making and incubates innovative algorithmic solutions.

The eventual collapse of desktop, web, mobile and **GIS SOFTWARE FUNCTIONALITY** into a seamless cross-platform ecosystem that encompasses geospatial widgets, wallpaper, smart and wearable gadgets will reinvent the ways that we interact with, access, analyze and query real-time data.

Our future **TRAINING, EDUCATION AND KNOWLEDGE TRANSFER IN LOCAL GOVERNMENT** curriculum will be an agile, real-time and transparent education model that is focused on capabilities and patterns of GIS use. It will allow professionals to learn, unlearn and relearn.

The future framework of **INFORMATION TECHNOLOGY (IT) INFRASTRUCTURE** is a highly secure hosted "as-a-service" cloud environment that will stimulate a ubiquitous GIS environment. The IT Department will focus on governance and management rather than implementation and maintenance.

FIGURE 3.2
The future of geospatial technology.

PILLAR 1   SMART GEOSPATIAL GOVERNANCE COMPONENTS

Geo-smart government will play a significant role in our future cities. The social democratization of **GIS Governance and Management Styles** will allow geospatial science to guide town, city and county growth so that it is aligned with individual citizen needs and public service delivery.

Today's Local Government Departments and Functions

Leadership is all about the vision, goals and objectives of an organization. Leadership is about influencing, motivating and enabling others to contribute to the organizations' success. Management is about controlling all of the moving geospatial parts to accomplish that goal. Governance and everything it entails has never been more important in local government.

Our objective is to describe the future components that any governance strategy must consider, embrace and successfully manage. The following is list of elements that will influence governance of the local government geospatial world:

1. The establishment of a Geographic Information Officer (GIO) in local government organizations.
2. A focus on civic engagement and management that includes a deeper approach to connecting with citizens and residents.
3. Managing an enterprise regional HUB solution.
4. Promoting a spatially literate society.
5. Promoting and managing open and transparent government through geospatial technology.
6. Building high-performance organizations (HPOs) with new performance metrics to enhance government operations.
7. Implementing Smart Business Practices (SBPs) for today's geospatial world.
8. Building a resilient community through good geospatial management practices.
9. Building a geospatially driven sustainable community through good management practices. This will include sustainable management regions, zones and neighborhoods.
10. Promoting and managing the regionalization of government operations through smart geospatial technology and geo-smart government.
11. Applying benefit realization planning and monitoring to geospatial technology.
12. Planning, designing and deploying geospatial fusion centers.
13. Improving sensitivity to a new culture with new values.
14. Managing advanced cyber security.
15. Improving advanced data interpretation techniques. Rapid real-time data will require understanding and interpretation.
16. Improving predictive analytics.
17. Focusing geospatial adoption across business lines.
18. Managing a total GIS ecosystem.
19. Managing education pathways throughout local government.
20. Managing the move to 3D GIS.
21. Advancing deeper geospatial capabilities across all local government departments.

22. Using 3D, web, mobile and desktop geo-ecosystem to map, manage and support decision related to the shifting challenges of the city of the future.
23. Understanding and managing the future challenges of local government.

The following list details the primary issues facing a local government GIS team. The GOI should be prepared and focused on ways to **map**, **manage**, **analyze**, **assess** and **use** geospatial technology to support the following future challenges:

- **Ecology**:
  - Improving the protecting wildlife natural habitat and natural resources
  - Reducing impact on the natural ecosystem
  - Managing and monitoring green and sustainable movements including green roofs, roof gardens, absorbent rain gardens, solar panels and small-scale farming within the future city
- **Preservation and Protection of Water Resources**:
  - The protection of water systems
  - Collection and cleansing of storm water
  - Improving water quality
  - Wetland restoration
  - Flooding and sea-level rise
  - Rainwater cleansing
  - Smart water and sensor technology to maximize irrigation
- **Renewable and Clean Energy**:
  - Urban buildings and the generation of energy
  - The production of energy within close proximity to the consumers
- **Waste and Recycling**:
  - Fully automated waste collection
  - Recycling and reuse
  - Automated recycling
- **Food and Urban Farming**:
  - Sustainable food generating practices
  - Improvements in food production, food delivery and disposal
  - Organic farming and organic farming standards
  - Local sourced food production

- **The Mobility of People**:
  - Safe, affordable and high-speed rail
  - Pedestrian space and walkable communities
  - Autonomous vehicles are key to our future
  - Road furniture and safety
- **Protecting Culture and Heritage**:
  - Promote and celebrate history and heritage
  - Recreation, arts, and community
  - Virtual and augmented reality
  - Cultural diversity
- **Community Livability**:
  - Accessibility and safety
  - Streamline access to nature, services and automated technology
  - Livability factors including amenities, demographics, economy, education, health care, housing, social capital, transportation and infrastructure
- **Public Safety and Hazard Mitigation, Emergency Planning, Response and Recovery**:
  - New urban policing strategies
  - Real-time incident data and decision-support technologies
  - Urban high rise monitoring and mapping
  - Hazard mitigation for natural and human-made hazards
- **Infrastructure**:
  - Future buildings
  - Water, soil and air quality
  - Designated pedestrian use areas
  - Track, measure and monitor infrastructure using sensor technology
- **Economy and Policy**:
  - Economic policy
  - Ecological sustainability
  - Resilient to economic downturns, hazards and demographic changes
- **Land Use and Zoning**:
  - The future of land use and zoning
  - New requirements about livability of cities, infrastructure, smart building codes
  - Marriage between industrial, agricultural, commercial and residential

- **Smart Buildings and Smart Technology:**
  - Smart buildings
  - Cross-purpose livable amenities
- **Self-Contained Resilient and Sustainable Neighborhoods, Regions or Zones:**
  - Sustainable neighborhoods index
- **Urbanization, the Urban Hub and the Mega City:**
  - The shape and form of cities as well as the urban lifestyle
  - Rail, road, natural boundaries and real estate are playing a significant role in the physical shape of cities

PILLAR 2   SMART DIGITAL DATA AND DATABASES COMPONENTS

Automated and universal collection, accuracy, standardization, reliability, access and effective use of **Digital Data and Databases** will become more important than software.

Advances in technology, procedures, protocols and interoperability continue to improve techniques for data collection, generation and analytics. These overarching trends create new data expectations and a new vision for data sharing that is open and transparent. This vision incorporates social media data, seamless business integration, improved data access and delivery, new ways of looking at the relationship of data and improved geospatial thinking and reasoning. The question for local government is: what is their vision of and expectations for data and databases? The fundamental smart practices that must be in place for all this advanced analysis is the primary data-related issue that needs addressing. Figure 3.3 reminds us that data can be fickle. We must constantly assess the content and quality of our data.

Digital data management and life cycle practices have never been more important. Today's geospatial world trends toward a centralized geospatial HUB of data, decisions, engagement, analysis, visualization and dissemination. This evolving data governance strategy relies heavily on the following smart data practices:

1. **Data Acquisition**: Smart real-time and innovative methods of capturing vast amounts of digital data are part of our local government landscape. Many classic and traditional data capture practices and principles support this assertion.

2. **Data and Databases Standards**: Data standards, format, definition and structure remain an important part of any local government GIS initiative. The standardization of digital data models is extremely important for organizational success. Esri's Local Government Information Model (LGIM) and the Next Generation 911 (NG911)

FIGURE 3.3
Tomorrow's data management life cycle practices.

standards, as well as Federal Damage Assessment data models, exemplify standards for local government achievement.

3. **Data Quality and Value Strategy**: The accuracy, validity, reliability, timeliness, relevance and completeness of data are more applicable now than ever. New technology and new tools used to collect

real-time data will become an even larger part of a GIO's day-to-day duties. The quality of geospatial data is critical to making informed and life-saving decisions.

4. **Data Analysis, Big Data Analytics and Geo-Enabled Processes and Actionable Analytics**: Data-based prediction of events and outcomes is on an upward trajectory. Dashboard statistical reporting practices require new modes of analyzing and presenting geospatial data. Bar charts, line diagrams, pie charts, histograms, scatter plots, dot plots, time series graphs and trend analysis will only gain importance. The three most important analytic procedures include:

   - Big Data Analytics
   - Applied Predictive Analytics and interactive patterns among data
   - Real-Time Data and Monitoring

   The abundance of digital data presents local government with information processing issues that exceed the capabilities of a traditional IT department. Supporting the use of information and understanding what data and databases to integrate into new systems are the real challenges here. The adoption of big data concepts and initiatives in local government varies widely. This is a product of education, training and innovative ideas about data relationships.

   Tools that translate increasing amounts of big government data into meaningful decision-support information are often the products of predictive analytics. In the future, they will include an examination of the science underlying algorithms and the principles and practices of predictive analysis. The importance of a local government's ability to successfully use predictive analysis techniques toward the effective interpretation and analysis of big data cannot be overstated.

5. **Data Distribution, Dissemination and Propagation**: The hallmark of a true enterprise, scalable and sustainable GIS model in local government is the successful and responsible dissemination of data using smart desktop, web, mobile and hosted virtualization tools, cloud technology and data mashup practices. Data "mashup" may be an old term, but you know what I mean.

6. **Data Maintenance and Upkeep**: The practices and protocols of managing, monitoring, updating and maintaining digital data layers are vital responsibilities for local government. Too many local governments have provided examples of revenue loss, loss of life and failed attempts to analyze GIS data due to their failure to keep digital data layers up to date and relevant.

7. **Data Visualization, Open and Transparent Data, and Community Managed Data**: For local governments, there is no getting around the challenges presented by data interpretation, representation,

symbolization and presentation. The move toward data visualization is relatively new. Only recently have local governments been introduced to new tools that provide techniques for presenting data in meaningful ways. Dashboards, story maps and analytical tools that find patterns within data are gaining significance. The craft of gathering from multiple data sources then communicating that information in a responsible and ethical manner will become a vital skill for local government GIS professionals.

We should also recognize that Open Data (OD) does not simply mean public data. It is any data that can be accessed through an Application Program Interface (API). It is data to be consumed by internal and external stakeholders. OD introduces the idea of mashing up different data sources to improve analytics and actionable decision support. OD also draws from data originating from the community. Citizen-managed data will play a significant role in our future.

Crowd-sourced data will prove increasingly important in the local governmental decision-making process while retaining its inherent challenges of latency, reliability, credibility and security.

> **Ideation**: A close cousin to the task of data visualization is something called "ideation." According to Wikipedia, ideation is the "creative process of generating, developing, and communicating new ideas, where an idea is understood as a basic element of thought that can be either visual, concrete, or abstract. Ideation comprises all stages of a thought cycle, from innovation, to development, to actualization."

Figure 3.3 details all of the components and complexity of the data management life cycle. Figure 3.4 graphically depicts the process involved in a data management strategy (DMS).

PILLAR 3   SMART PROCEDURES, WORKFLOW AND INTEGRATION OR INTEROPERABILITY

The Smart City HUB will focus on optimized and automated **procedures, workflow and interoperability** as well as an **agnostic System of Systems** that performs evidence-based analysis, supports data-driven decision making and incubates innovative algorithmic solutions.

There is no doubt that any successful future smart government initiatives will require real-time data and meaningful information obtained from multiple internal government software solutions and external open data sources. Successful future governments will require agnostic system of systems that talk and communicate with each other, allowing for total interoperability and

FIGURE 3.4
Data management strategy (DMS).

sophisticated decision-support procedures, processes and algorithmic solutions. As I have stated, the smart city HUB will focus on optimum automated procedures, workflow and interoperability. It will require an agnostic system of systems that performs evidence-based analysis and data-driven decision making.

As the IoT grows in tandem with the miniaturized wireless technology market *within* the operations and functions of local government it may, **without proper procedures, workflow, protocols and standards become increasingly disjointed and potentially disorganized**. Technology in our vehicles, clothes (wearables), security systems, education and other systems in our lives that connect to the cloud and provide us with information feedback are all components of the new technological ecosystem of local government. The challenges we face with interoperability stem from connectivity, reliability and security between devices along with the differences between applications, platforms and standards.

The pieces of this puzzle take the forms of new procedures, new workflows and the complete interoperability of disparate hardware, software and database systems. The future demands platform, device and system agnostic approaches to the local government technology ecosystem. Agnostic – in an IT and GIS context – refers to a solution that is interoperable among various systems. It denotes the ability of something to function without knowing the underlying details of a system that it is working within. The term also encompasses the ability of computer systems and software to utilize and share information between devices regardless of their manufacturer. Device **agnostic** is the capacity of a computing component to work with various **systems** without requiring any special adaptations.

We must therefore understand each of the following:

- Geospatial procedures, protocols and workflows
- Geospatial interoperability
- Agnostic system of systems
- Enterprise geospatial regional hubs

Interoperable Systems in Local Government

The amount and interrelationship of data and databases in local governmental departments are of the utmost importance to the system of systems world that will find us in the near future. If we look at each local government department and corresponding database in addition to the contemporary goals and objectives of government, we can see the clear importance of successfully integrating different databases. With that in mind, let's look at a number of unique databases that already exist within local government.

Local Government Departments and Software and Databases Integration Opportunities

- **Public Safety and Law Enforcement**:
  - Police
  - Sheriff
  - Emergency Management
  - EOC
  - Fire Department
  - Animal Control
- **Public Works and Public Utilities**:
  - Water
  - Sewer
  - Storm Water
  - Solid Waste
  - Recycling
  - Engineering
  - Transportation
  - Electric
  - Telecommunications
- **Land and Information Management**:
  - Planning and Zoning Department
  - Economic Development

- Building and Inspections
- Code Enforcement
- Tax Assessor
- Information Technology Department
- Public Information Officer (PIO)
- **Natural Resources, Parks and Recreation**:
  - Parks and Recreation
  - Tree Management and Arborist
  - Environmental and Conservation
  - Cooperative Extension
- **Public Administration**:
  - Executive Management
  - Legal Department
  - Finance Department
  - Housing Department
  - Environmental Affairs
  - Elections
- **Public Services**:
  - Library
  - Schools
  - Public Health
  - Social Services
  - Community Development
- **Telecommunications**:
  - Local Government Telecommunications Broadband Service Providers

Alignment of a Local Government Vision, Goals and Objectives

The vision, goals and objectives of a local government must be reconciled with the data and databases supporting the vision.
  #### The Challenge Statement for Local Government

- Feel safe and secure
- Earn a good living
- Move around my community
- Enjoy a healthy environment
- Be empowered and included in society
- Live an active and healthy life

Figure 3.5 illustrates the existing corporate silo approach to database management. Historically, local government have made little attempt to integrate disparate, dissimilar, unrelated and departmental databases. The figure illustrates how the integration of each departmental database would further organizational goals.

Figure 3.5 illustrates departmental databases as components of the total working database structure. It shows the ways that local government databases must be integrated and used in conjunction with one another.

PILLAR 4   SMART GEOSPATIAL SOFTWARE COMPONENTS

The eventual collapse of desktop, web, mobile and **GIS Software Functionality** into a seamless cross-platform ecosystem that encompasses geospatial widgets, wallpaper, smart and wearable gadgets will reinvent the ways that we interact with, access, analyze and query real-time data.

Though we see a clear line between desktop, web and mobile GIS software solutions today, our future smart GIS ecosystem may be focused on cross-platform business solutions in different forms. The line between the functionality of desktop solutions, web browsers, and mobile tools is slowly disappearing. Accordingly, we must embrace cloud-based business applications that can run on any operating system.

As a jumping off point, let's discuss current smart geospatial software.

- **Desktop GIS**: Desktop solutions offer the most sophisticated geospatial functionality and the richest user experience. Desktop solutions are arguably the most responsive. The primary downside to desktop applications is that the end user must download then install a software application, as opposed to utilizing a browser-based solution. This reality increases cost and decreases practicality for local government professionals who want to immediately engage with a solution. Oftentimes, basic cyber security programs prevent the download of desktop tools.

- Desktop applications pose another potential problem are: the potential constraints imposed by a desktop's Operating System (OS) platform. OS platforms include Windows (Visual Studio), Mac (Xcode) or Linux (Eclipse). Each platform brings completely different set of commands, procedures, protocols and Application Programming Interfaces (APIs) to the table.

- **Web Application**: Web applications allow users to access software tools instantly and directly through their existing browser. It requires no download and installation process. The challenges for the web application are responsiveness (especially if data is used from the

DEPARTMENTAL DATABASES

Vision, Goals and Objectives Local Government	Group 1: Heavy Hitters	Group 2: Technical Specialists	Group 3 – Easy Enthusiasts	Group 4 – The Outliers
Feel safe and secure	• Police Department • Sheriff Department • Geographic Information Systems (GIS) Division	• Emergency Management and Emergency Operations Center (EOC) • Fire Department	• Animal Control • Public Information Officer • Parks and Recreation Departments	• Legal Department • Public Health
Earn a good living	• Geographic Information Systems (GIS) Division	• Economic Development • Transportation	• Public Information Officer • Community Development	• Housing Department • Schools • Telecommunications
Move around my community	• Geographic Information Systems (GIS) Division	• Economic Development • Transportation	• Public Information Officer • Parks and Recreation Departments • Community Development	• Housing Department • Libraries • Social Services • Telecommunications
Enjoy a healthy environment	• Water, Sewer, Storm water, Gas and Electric • Information Technology (IT) Department • Geographic Information Systems (GIS) Division	• Planning and Zoning Department • Electric Department • Economic Development • Transportation	• Animal Control • Building and Inspections • Code Enforcement • Public Information Officer • Parks and Recreation Departments • Tree Management and Arborist • Environmental and Conservation • Cooperative Extension • Community Development • Solid Waste and Recycling	• Housing Department • Environmental Affairs • Libraries • Schools • Social Services • Public Health
Be empowered and included in society	• Police Department • Sheriff Department • Information Technology (IT) Department • Geographic Information Systems (GIS) Division	• Planning and Zoning Department • Economic Development • Transportation	• Code Enforcement • Public Information Officer • Parks and Recreation Departments • Environmental and Conservation • Community Development	• Environmental Affairs • Telecommunications
Live an active and healthy life	• Geographic Information Systems (GIS) Division	• Economic Development • Transportation	• Parks and Recreation Departments • Environmental and Conservation • Community Development	• Environmental Affairs • Libraries • Schools • Social Services • Public Health • Telecommunications

GEOGRAPHIC TECHNOLOGIES GROUP

FIGURE 3.5
Departmental databases and the vision, goals and objectives of local government.

server), web-related security concerns (security of a web server is a significant undertaking) and performance. The development of web applications includes the user interface and the server aspects of the solution. Three applications for the client include HTML, JavaScript and CSS. AJAX and PHP are applications for the server. Building a web application requires a deep understanding of the multiple technologies intersecting between the server side and client side.

- **Mobile Applications**: As the name implies, mobile application can be used anywhere, anytime. Mobile solutions are an integral component of the smart city initiative. These applications run locally on user devices and are generally fast and responsive. Downloading and updating software tailored to a specific OS tends be much easier than the alternative. The challenge comes with planning and designing layouts for small screens. Screen size limits user experience and persists as a defining factor in the development of mobile applications. Like its desktop-based counterpart, mobile applications present OS, API's and language challenges. Mobile application developers often encounter issues similar to their counterparts working in the desktop application development space.

- **Standard Seamless Cross-Platform Ecosystem of Tools**: I anticipate an eventual collapse of desktop, web and mobile **GIS Software Functionality** into a singular and seamless cross-platform ecosystem of geospatial widgets, wallpaper, smart and wearable gadgets.

Software Functionality Initiatives

The public participation software solution is yet another smart geospatial software initiative that will play a prominent future role in smart cities. How will this new technology shift of participatory GIS, public engagement, open data portals, and crowdsourcing solutions fit into the future of local government? Today, the practice of participatory GIS is engineered for community empowerment through user-friendly, measured, demand-driven, and integrated cloud-based geospatial solutions. In the future, we will see cross-platform approaches to participatory GIS.

Smart geospatial software of the future will likely include the following characteristics:

- The collapse of geospatial web, mobile and desktop **GIS software functionality** into a standard seamless cross-platform ecosystem of geospatial widgets, wallpaper, smart and wearable gadgets that will reinvent how we interact, access, analyze and query real-time data.

- Geospatial solutions will be used extensively for outreach, engagement, interaction and feedback from community residents.

- Crowdsourcing solutions will continue to play a significant role in our local government smart city initiatives.
- GIS-based public and civic engagement applications will continue to improve public opinions about government due to their integrative approaches to shareholder participation.
- Civic management will focus on the ways geospatial technology like usable crowdsourcing solutions, external agency interfaces, bi-directional applications, and open data portals with instant real-time feedback mechanisms will positively affect local government.
- Three-dimensional (3D) GIS will become a standard practice. The City of Mississauga, Ontario, and many other local government organizations are already demonstrating the ways 3D GIS tools will play a role in the future local governments.
- Indoor GPS and indoor smart GIS mapping will become common-place in the software market. Local governments will likely be slow to embrace specifically tailored indoor GPS systems.
- The development of bespoke geospatial design solutions will become local government initiatives. "Bespoke" describes a product that has been commissioned by a specific client and customized to their needs. In the near future, the following local governmental departments will require bespoke geospatial design solutions:
  - Public Safety and Law Enforcement
  - Emergency Management
  - Emergency Operations Center (EOC)
  - Fire Department
  - Public Works and Public Utilities
  - Engineering
  - Transportation
  - Electric
  - Telecommunications
  - Land and Information Management
  - Economic Development
  - Building and Inspections and Code Enforcement
  - Tax Assessor
  - Public Information Officer (PIO)
  - Parks and Recreation & Environmental and Conservation
  - Cooperative Extension
  - Public Administration
  - Executive Management

- Legal Department
- Finance Department
- Community Development and Environmental Affairs
- Elections
- Public Services
- Public Health
- Social Services
- Local Government Telecommunications:
- The enterprise regional HUB **solution will play a significant role in the world of GIS in local government**.
- **Gamification** is the application of game-design elements and principles in non-game contexts. Gamification will be used ever more frequently by local government to promote more meaningful interfacing amongst citizens and employees. Gamification for government services, applications and processes increases user interactivity and changes behavior to the benefit of engagement and efficiency. Citizens or employees who have fun are more likely to change their behaviors.

PILLAR 5   SMART TRAINING, EDUCATION AND KNOWLEDGE TRANSFER COMPONENTS

Our future **Training, Education and Knowledge Transfer in Local Government** curriculum will be an agile, real-time and transparent education model that is focused on capabilities and patterns of GIS use. It will allow professionals to learn, unlearn and relearn.

One of the most important components of a successful smart city and enterprise GIS initiative is easy to ignore: the training. It is the training, education and transfer of knowledge strategies deployed by a local government that will play significant and long-lasting roles in smart city initiatives. Local government is becoming more business-centric and demonstrating a slow but noticeable shift in the mindset, inventiveness, resourcefulness, ingenuity, and priorities toward an alignment with the smart community initiative. Corresponding changes in the ways governments look at geospatial technology will parallel this shift. More and more towns, cities and counties are considering re-engineering the role of geospatial technology in favor of a smart city solution.

Additionally, the way geospatial technology is used for future geo-smart government priorities will shift. A workforce prepared for the challenges of geospatial smart technology is critical to the success of any organization. An optimum governance model with clear lines of responsibility, communication and structured accountability coupled with workforce management and engagement are characteristics of a truly smart enterprise geospatial solution. A governance strategy that facilitates on-going research and development of new geospatial technologies buttresses an organization with resilience and sustainability.

Governments that fail to adjust their geospatial learning management practices will struggle with organizational growth and productivity. As a result, we will see local governments abandon traditional methods of learning and knowledge transfer in favor of more effective contemporary solutions.

Up-to-date training, education and knowledge transfer practices applicable to smart cities include **e-learning, instructor led and in-person training and education, virtual training and education and knowledge transfer solutions, self-study strategies, on-the-job training, video and animation solutions, and the support of performance monitoring and measurement technology.**

In order to sustain businesses operations, functions and efficiencies with local government, GIS staff need to continuously adapt. The ability to learn new information anywhere, at any time is a critical part of the future landscape. The five following components are critical to an organization's future success:

- Learning agility
- Customized learning that's new and interesting
- Learning culture
- Collaborative learning
- Tools and technology to monitor and measure

The key factors that will play significant roles in the training, education, and knowledge transfer among a smart city's GIS professionals are:

1. **Geospatial performance, competency, ability and results are the leading drivers** for gauging the effectiveness of a local government organization's training, education, knowledge transfer and overall learning programs.

2. **Some management groups indicate that what you know is less valuable than what you can do.** Sharing your knowledge is far more important than hoarding it to seem like an expert. In a performance-based learning organization, *the team* is the expert.

3. **Local government organizations must plan to focus heavily on learning delivery that appeals to a millennial workforce.** Millennials understand technology for training, education and transfer of knowledge to be ubiquitous, available and omnipresent. This maturing global demographic strongly prefers technology-based and personalized learning methods. For millennials, learning

is not the event. Learning is the starting line. At work, millennials want downstream support that helps them adopt and adapt to their new knowledge. This cultural shift is yet another reason for the rise of cloud-based, collaborative and social learning-oriented support tools that are available anytime and anywhere. A "Learning Organization" spans the entire enterprise and provides real, measurable value for an organization.

4. **Maintaining optimal performance management for organizations requires a Continuous Learning Environment (CLE).** It must include Blended Learning. Blended Learning refers to a variety of customized learning options that best meet the needs of the learners. **Knowledge transfer is a part of maintaining optimal performance management for organizations.** A formal knowledge transfer process is vital to create more effective training experiences, keep the local government running efficiently and streamline succession planning.

PILLAR 6 SMART INFORMATION TECHNOLOGY (IT) INFRASTRUCTURE COMPONENTS

The future framework of **Information Technology (IT) Infrastructure** is a highly secure hosted "as-a-service" cloud environment that will stimulate a ubiquitous GIS environment. The IT Department will focus on governance and management rather than implementation and maintenance.

To stay relevant in the technology-driven managerial environment of towns, cities or counties local governments need robust and flexible IT infrastructure that can quickly, efficiently and securely respond to the emerging requirements of local operations and functions. Most IT departments seek such a solution yet remain adverse to the overhead and complexity that characterize implementation and maintenance. Historically, the IT department has done all the heavy lifting of selecting, implementing, and maintaining the infrastructure that backbones an IT ecosystem. The question we have to ask is: what are the mission critical IT infrastructure and technological priorities for a local government organization today and in the near future?

Currently, the standard IT infrastructure includes but is not limited to the following components:

- **Hardware**: Servers, computers, data centers, switches, hubs and routers and other equipment
- **Applications and Software**: Enterprise resource management (ERM), customer relationship management (CRM), and productivity software applications

- **Networking**: Internet connectivity, firewalls, load balancers and security systems
- **Data Storage**: Network attached storage and storage area network (SAN) enabled storage. Enterprise data warehouse, information technology infrastructure library (ITIL) and rightsizing
- **Facilities**: The physical data centers, power, cooling and security components and data center infrastructure management (DCIM)
- **Governance and IT Policy**: Professional IT staff, network administrators (NA), developers, designers and end users with access to any IT infrastructure, and Enterprise Architecture (EA)
- **Service Management**: The management of infrastructure as a service (IaaS) and software as a service (SaaS)

IT infrastructure is the strategic amalgamation of hardware, software, network and communications infrastructure needed to handle technology workloads for the efficient deployment, operation, management and smooth running of a local government organization. The array of multi-disciplinary obstacles facing IT management and governance includes, but is not limited to the following array of technical challenges:

- Cyber Security: Privacy and Security
- Real-Time Data
- Data Acquisition
- Virtualization
- Data Analytics
- Cloud Computing
- Mobile Devices
- A Connected Smart City Ecosystem: A System of Systems
- Agnostic Technology Platform
- Smart Connected Assets and Systems
- Smart Devices: Sensors and Cameras
- Cloud Software and Cloud GIS
- Citizens as Sensors
- Wearable GIS
- New IoT Platform
- Driverless Cars
- Internet Glasses and Contact Lenses
- Micro Chips
- Computers – Wall Screens, Flexible Electronic Paper and Virtual Worlds

- Inexpensive Miniature Devices and Sensors
- Internet of Things (IoT)
- Unmanned Aircraft Systems (UAS) and Drone Technology
- Mobile Solutions
- Expanding Wireless and Web Networks
- Advances in Computer Capacity and Speed
- Data Collection and Data Generation
- Big Data Analytics
- Data Models
- Improved Communication Tools
- New Software Applications

IT infrastructure environment of local government massively influences their utilization of geospatial technology. Local governments are evolving, progressing and changing the way they embrace IT infrastructure. Outsourcing IT services is already commonplace. The term "as-a-service" is a relatively new term. Google explains that "as-a-service" is "a collective **term** that refers to the delivery of anything **as a service**. It recognizes the vast number of products, tools and technologies that vendors now deliver to users **as a service** over a network – typically the internet – rather than provide locally or on-site within an enterprise." In the near future, this phenomenon could be termed "aaSS," an abbreviation for "as-a-Service" system. Figure 3.6 illustrates the evolving IT outsourcing of "as-a-service" models in local government.

New and emerging technological trends and the worldwide smart city – that system of systems – both require robust IT Infrastructure and an expert IT staff. The IT challenges of the future require new levels of expertise in software, hardware, communications, programming, security and more. Where will this expertise be located?

As local government become ever more dependent on cloud platforms and outsourced services, the role of IT continues to change. We have seen the Internet expand beyond personal computers (PC) and mobile devices into enterprise assets such as field equipment, sensors, and video technology. Smart city plans explore the possibility of processing the large amounts of data coming into the IT Department from video cameras, parking sensors, air quality monitors, rover gauges and traffic monitoring devices in order to help local governments achieve their goals.

Large IT departments with deep pockets and abundant resources are funding progressive IT initiatives that focus on cloud platforms and outsourced services. Additionally, evidence indicates that mid-sized local government organizations are already creating cost-effective, secure and responsive hybrid infrastructure solutions that scale and adapt to the business needs of every department without the investment of completely in-house

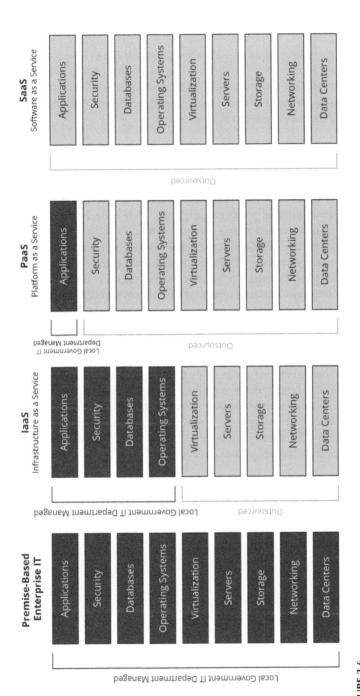

FIGURE 3.6
Evolving IT outsourcing "as-a-service" models in local government.

infrastructure. Software as a service (SaaS), outsourced IT operations and cloud-based or third-party managed infrastructures are taking over the day-to-day operations of local government. Logically following this shift to externally manage and maintain IT, local governmental IT departments are becoming more technology governance groups, less technological implementation groups. The main reason local governments have pushed cloud adoption stem from its cost reduction, speed of procurement and deployment and its responsiveness to regulations.

A More Ubiquitous GIS Environment

New IT Infrastructure models promote a ubiquitous GIS environment. Even though web-based GIS solutions are not a new architectural phenomenon, they remain the essential modern solution. Web-based GIS solutions are deployed and implemented using a cloud ArcGIS Online model, or an "on premise" model using ArcGIS Server or – as is the case with many local government organizations – a hybrid combination that leverages the best of both "on premise" and cloud worlds. Esri's new portal is a key component of the modern enterprise, scalable and sustainable smart GIS world. It provides a framework for sharing and utilizing maps, apps and data. It's important to note that the shift from client/server architectures to web services enables users to connect to a plethora of information, from enterprise databases to the Internet of Things (IoT), sensors, participatory data and big data analytical tools. When architecture pushes GIS from static data mode into real-time mode, the number of users and uses inevitably shifts. According to Esri, web-based GIS allows users to move with greater agility from custom application development to configurable templates and web app builders. Most importantly, web GIS moves us from proprietary data into open data and shared services that can empower everyone.

4

Performance: High-Performance Organizations (HPO)

It is not the strongest of the species that survives, nor the most intelligent that survives. It is the one that is most adaptable to change.

– Charles Darwin

The ability that will prove most essential to our success is no longer the technical, classroom-taught left brain skills. Instead, our greatest advantage lies in what we humans are most powerfully driven to do for and within another, arising from our deepest, most essentially human abilities – empathy, creativity, social sensitivity, storytelling, humor, building relationships, and expressing ourselves with greater power than logic can ever achieve.

– Geoff Colvin

Introduction

Since the inception of local government, experts have shaped, reshaped and defined management styles, theories and strategies. Notables characters in the history of management science include Taylor, Gantt, Gilbreth (s), Fayol, Weber, Follett, Barnard and the team behind the Hawthorn Studies. In today's world, ever increasing demands from internal departments, external residents and citizen stakeholders force local government to explore new ways of improving performance and enjoying organizational success.

Geospatial technology presents more management challenges than any other local governmental discipline. As I stated in 2017, "the outcome of an ill-defined, haphazard governance model leads to poor GIS service delivery and an unsustainable management strategy" (Holdstock, 2017). Furthermore, any GIS department endeavoring to deal with its abundant data, workflow complexity, people, software, technology and politics dooms itself to failure if it lacks a management model that promotes quality, open, and continuous improvement. For this very reason, the high-performance organization (HPO) strategy is a core component for future success within the geospatial and geo-smart government framework.

The objective of this chapter is to introduce the concept and key characteristics of HPO in order to facilitate improvement and sustainable performance of geospatial management operations in local government. This chapter aims to provide insight into optimum organizational design, structure, process, technology, leadership, people, culture and external stakeholders.

The HPO is an alternative framework model that, according to most literature and consulting companies, leads to "improved, sustainable organizational performance." Based on my interest, research and exposure to local government HPOs, I concur with researchers who posit that there is no absolutely clear definition of HPOs. HPOs are about good management, action orientation, openness, a long-term drive, a continuous improvement mindset and workforce quality.

Below are a few excellent definitions of HPOs:

> A High Performance Organization is an organization that achieves financial results that are better than those of its peer group over a longer period of time, by being able to adapt well to change and react to these quickly, by managing for the long term, by setting up an integrated and aligned management structure, by continuously improving its core capabilities, and by truly treating the employees as its main asset.The Characteristics of a high-performance organization (HPO)
>
> **– Andre A. de Waal.**

> Until now there has been no generally accepted name or definition of HPO's, and in the literature the HPO is often referred to as the accountable organization,, the adaptive enterprise, the agile corporation, the flexible organization, the high performance work organization, the high performance work system, the high reliability organization, the intelligent enterprise, the real-time enterprise, the resilient organization, the responsive organization, the robust organization, and the sustainable organization.The Secret of High Performance
>
> **– Andre A. de. Waal.**

> The High Performance Organization (HPO) Framework is a conceptual, scientifically validated structure that managers can use for deciding what to focus on in order to improve organizational performance and make it sustainable. The HPO Framework isn't a set of instructions or a recipe that can be followed blindly. Rather it is a framework that has to be translated by managers to their specific organizational situation in their current time, by designing a specific variant of the framework fit for their organization.
>
> **– Google**

> Sustained high performance is always the goal, but many managers lack the tools to accomplish it The High-Performance Organization (HPO) Diagnostic model asks managers to consider three conceptual change levers (leadership, vision, and values). And three applied change levers (strategy, structure, and systems). Collectively these levers offer a

> systematic approach to sustained high performance. Making these lev-
> els work requires addressing seven diagnostic questions" Building High
> Performance Organizations, Pickering and Brokaw.

There is an abundance of experts advertising their skills and services regarding this new HPO strategy. For example, the experts at the HPO Center (https://www.hpocenter.com/) work with numerous global organizations to improve long-term performance. They offer lectures, workshops, Master Classes and training on topics such as: high-performance organizations, high-performance leadership, quality of management, HPO and best ideas. Their website provides a wonderful explanation of HPO theory and is a quality source for information about local government GIS programs.

The International City/County Management Association (ICMA) is an organization of local government professionals dedicated to creating and sustaining thriving communities throughout the world. ICMA has identified the high-performance organization model as a leading organizational development practice for local governments. In partnership with the *Commonwealth Centers for High-Performance Organizations* (CCHPO), the ICMA delivers the education and technical assistance required to implement these methods in an organization.

The ICMA and Commonwealth Centers offer eight strategic HPO training modules, including:

- Module I: Setting the context
- Module II: Overview of HPO Diagnostic/Change model that includes the six change levers
- Module III: Leadership Philosophy
- Module IV: Leadership Function
- Module V: Leadership Form
- Module VI: Where work need to be cone in an Organization
- Module VII: The Vision to Performance Spiral
- Module VIII: The Values to Work Culture Spiral

The document supplied by CCHPO is over 250 pages and includes multiple questionnaires and numerous evaluative strategies.

My introduction to HPOs came in the midst of a GIS strategic planning project for the City of Boulder, Open Space and Mountain Parks Department (OSMP). OSMP management was interested in the ways they could combine the methodology of geospatial strategic planning with the strategies of a high-performance organization. I was introduced to Pickering and Brokaw's diagnostic change model strategy for HPOs, titled *Building High-Performance Organizations (1993)*.

At that time, I assumed that HPO was a recognized management model, formula, procedure, or recipe created by a specific strategist or management theorist. I learned – to my surprise – that HPO strategy is synthesized from the ideas, thoughts, approaches and tactics outlined by researchers and professionals in well-known publications. To my delight, I realized that I had actually been reading some of the books that built HPO. These works included *Good to Great* (Collins), *The High Performance Organization* (Holbeche) and *What Makes a High Performance Organization* (De Waal). Other books that further the HPO philosophy include *Good Leaders ask Great Questions* by John C. Maxwell (2016).

What Makes a High Performing Organization by De Waal is an excellent book for all GIOs and GIS coordinators, managers and specialists working in local government. De Waal's extensive understanding of HPO strategies stem from his in-depth evaluation of HPO studies.

What Makes a High Performing Organization?

What can GIS Managers and GIOs take from De Waal's work? How can they use that knowledge to create an HPO? Consider the below list of key points – drawn directly from HPO specialists – as a launchpad.

- Good leadership
- Hiring the right people
- Understanding threats to the organization
- Discipline
- Technology accelerators
- Perseverance
- Change management
- Innovation
- Reaching maximum potential
- Organization encouragement
- Structure
- Making the organization a great place to work
- Creating a value-based organization
- Good management
- Empowerment and accountability
- Strategy
- Establishing an internal structure

- Available resources
- Incentives
- Alertness
- Agility
- Adaptiveness
- Alignment

Figure 4.1 illustrates what makes an HPO according to De Waal.

Understanding that HPOs are defined by a set of important characteristics is of critical importance. An HPO does not come equipped with a cookie cutter methodology. The HPO framework does not provide a blueprint for organizations to follow. A successful HPO framework requires an understanding of organizational experience, expertise and creativity. At the same time, this framework must transform an organization into an HPO. Moral of the story is: for a management strategy to succeed, it must evolve and reflect the changing circumstances of the times.

Some additional characteristics of HPOs include the following:

- Adaption and change
- Goals setting
- Flexible
- Customer focused
- Team spirit and team work
- Investment in workforce
- Flatter hierarchies
- Cross functional collaboration
- Diversity
- Continuous improvement

The following are key points that have come from HPO Specialists.			
Good Leadership	Hiring the Right People	Understanding Threats to the Organization	Discipline
Technology Accelerators	Persaverance	Change Management	Innovation
Reaching Maximum Potential	Organizational Encouragement	Structure	Making the Organization a Great Place to Work
Creating a Value-Based Organization	Good Management	Empowerment and Accountability	Strategy
Establishing an Internal Structure	Available Resources	Incentives	Alertness
Agility	Adaptiveness	Alignment	

FIGURE 4.1
What makes a high performing organization – According to de Waal.

- Core capabilities
- High commitment organizations
- Trust
- Integrity
- Role model
- Coaching
- Achieving results
- Confidence
- Decisive about non-performers
- Responsibility
- Trusted management

For an example of the City of Grants Pass, Oregon's commitment to HPO strategy: https://www.grantspassoregon.gov/1080/High-Performance-Organization-HPO.

The Five Success Factors of an HPO

Here are five HPO factors:

1. Quality of management
   - Belief and trust in others
   - Fair treatment of others
   - Team members have integrity
   - Decisive
   - Action orientated
   - Holds people accountable
   - Openness
2. Open culture
   - Open culture
   - Team members opinions are valued
   - Trial and error strategy
   - Knowledge exchange
   - Learning

- Communicating
- Performance driven

3. Long-term orientation
   - Partnerships
   - Stakeholders
   - Relationships
   - Internal hiring
   - Encouragement to be managers

4. Continuous improvement and renewal
   - Continuous improvement
   - Simplifying
   - Innovation
   - New strategies
   - Competitive advantage (local government)
   - Improved efficiency
   - Core competencies

5. Quality of employees
   - Complimentary team members
   - Diverse group
   - Encouraged to develop skills
   - Accountable
   - Responsible
   - Creativity higher levels of effectiveness

Figure 4.2 illustrates the core components of an HPO.

FIGURE 4.2
Five success factors of a high-performance organization (HPO).

The Nine Questions an HPO Asks

Figure 4.3 illustrates the nine questions an HPO asks.

Supporting the prior nine questions are seven diagnostic questions that are fundamental to the success of an organization.

There are seven key diagnostic questions that are fundamental to success.

1. What is high performance?
2. How can a local government organization know if it is high performing?
3. According to whom is an organization high performing?
4. Why does an organization need to be high performing?
5. Is the organization doing the right thing?
6. How good is an organization at doing the right thing?
7. How should the organization treat each other including partners/customers other stakeholders?

Figure 4.4 is an example of a city department that has answered the nine key diagnostic questions important to the success of an HPO.

Continuous improvement is a major factor in HPO. The following six steps define strategies for continuous improvement within organizations:

1. **Improvement based on small changes**: Improvements are based on small, organic changes rather than larger radical changes that might arise from research and development.
2. **Staff ideas increase the rate of adoption**: If ideas come from the workers themselves, they are less likely to be radically different and therefore easier to implement.

The 9 Questions a High Performance Organization Asks
1. Who currently are our stakeholders, beneficiaries, customers, partners, collaborators, suppliers, and others?
2. Who should our stakeholders or potential stakeholders be?
3. What do each of our stakeholders or potential stakeholder's value? What do they wants, needs, or, and expect?
4. What does high performance mean to us?
5. How does our vision fit within that of the organization?
6. How would we describe the desired future state we are seeking?
7. How would we know if we were high performing? How would we measure it?
8. How do we define quality for our products and services?
9. How are we going to treat each other and our stakeholders?

FIGURE 4.3
The nine questions an HPO asks.

High Performance Organization Strategy
9 Key Diagnostic Questions (KDQ)
City of Boulder, Open Space and Mountain Parks

Open Space and Mountain Parks Department Vision

- Treat others as valued partners
- Develop feedback loop to OSMP
- Environment is critical
- The Key Diagnostic Questions (KDQ) and Change Levers of an HPO are critical to success
- The City , RIS and OSMP must have a shared vision
- The HPO "Vision to Performance Spiral" is a useful tool (strategic thinking, strategic planning, tactical operations planning and execution, and monitoring and recovery)

1. **Who currently are our stakeholders including beneficiaries, customers, partners, collaborators, suppliers, and others?**
   - All OSMP workgroups
   - Public
   - City Departments
   - Boulder County
   - Private Consultants
   - Jefferson County
   - Other Governmental Agencies
   - FEMA
   - State of Colorado
   - Appointed Officials

2. **Who should our stakeholders be? Who are our potential stakeholders?**
   - All OSMP Workgroups
   - Public
   - City Departments
   - Boulder County
   - Private Consultants
   - Jefferson County
   - Other Governmental Agencies
   - FEMA
   - State of Colorado
   - Appointed Officials
   - EOC

3. **What do each of our stakeholders or potential stakeholder's value? What are there wants, needs, and expectations?**
   - Access to data and databases
   - Open data portals
   - Published map services
   - Hardcopy maps and cartographic standards (~1,000 maps created annually)
   - Data, data analytics, and statistics
   - Data inventory and data collection support
   - Training and Education
   - Web tools and web exposure
   - Immediate response
   - Customer configured solutions
   - Enabling stakeholders

4. **How does our vision "align" within that of the city and OSMP?**
   - City Vision: Service excellence for an inspired future RIS is moving towards aligning OSMP and the city's vision statement into their culture

5. **How would we describe the desired future state we are seeking?**
   - Calm working atmosphere
   - Proactive rather than reactive
   - Enabled users rather than dependent
   - Maintaining tools that enable users rather than receiving work requests from them
   - Efficient Process

6. **What does high performance mean to us?**
   - Reliability
   - Innovation
   - Quality Products
   - Efficiency
   - Timely response
   - Automating manual procedures
   - Collaboration
   - Culture
   - Vision
   - Customer service

7. **How would we know if we were high performing (metrics)?**
   - Increased self-sufficiency of stakeholders (i.e. 5% reduction in work requests for a printed map by end of Q2
   - Monitoring Key Performance Indicators on an annual basis
   - Annual stakeholder survey
   - Decreased front desk contact with stakeholders and public (i.e. 5% reduction in front desk contacts by 2019)
   - Engaging the public more (i.e. add two new community activities to 2018 calendar)
   - Publishing papers (i.e. publish two articles in 2018)
   - Winning special achievement (SAG) award

8. **How do we define quality for our products and services?**
   - Metadata – data about data
   - Adoption rate of QUALITY products
   - Direct feedback
   - Surveys

9. **How are we going to treat each other and our stakeholders?**
   - Respect
   - Integrity
   - True partners

FIGURE 4.4
High-performance organization strategy: City of Boulder, Open Space and Mountain Parks.

3. **Small improvements are cost effective**: Small improvements are less likely to require major capital investment than major process changes.

4. **Working knowledge and talent**: The ideas come from the talents of the existing workforce, as opposed to using research, consultants or equipment – any of which could prove very expensive.

Improvement Based on Small Changes	Improvements are based on many small changes rather than the radical changes that might arise from research and development.
Staff Ideas Increase the Rate of Adoption	As the ideas come from the workers themselves, they are less likely to be radically different, and therefore easier to implement.
Small Improvements are Cost-Effective	Small improvements are less likely to require major capital investment than major process changes.
Working Knowledge and Talent	The ideas come from the talents of the existing workforce, as opposed to using research, consultants, or equipment – any of which could be very expensive.
Involves Everyone in the Organization	All employees should continually be seeking ways to improve their own performance.
Staff Take Ownership	It helps encourage workers to take ownership for their work, and can help reinforce team working, thereby improving worker motivation.

FIGURE 4.5
The Six steps that define continuous improvement within an HPO.

5. **Involves everyone in the organization**: All employees should continually seek ways to improve their own performance.

6. **Staff take ownership**: When workers take ownership of their results, it promotes team work, accountability and motivation.

Figure 4.5 shows the six steps that define continuous improvement within an HPO.

A High-Performance Geospatial Organization (HPGO)

How can the management strategy of an HPO be used for geospatial technology in local government? The HPO is a conceptual, scientifically validated structure that managers utilize when deciding what to focus on in order to improve organizational performance and sustainability. Though tailored to organizational needs, the HPO is a recipe for success.

The concept and key characteristics of HPOs can facilitate improvement and sustain performance in geospatial management and operations in local government. My research on leading HPO experts offers some insight into optimized organizational design, structure, process, technology, leadership, people, culture and external stakeholders.

Local government GIS programs are complex and multifarious. Prior to introducing HPO strategy to an organization, one must recognize the following characteristics typical of GIS programs (Figure 4.6):

1. A Ratified GIS Vision, Goals and Objectives
2. A Strategic, Enterprise, Sustainable, Scalable, Enduring GIS
3. An Optimum GIS Governance Model – Support and Operations
4. Key Performance Indicators (KPI)
5. A High-Performance Organization (HPO) – Quality Mindset

Characteristics of a High Performance Geospatial Organization (HPGO)
1. Ratified GIS Vision, Goals and Objectives
2. A Strategic, Enterprise, Sustainable, Scalable, Enduring GIS
3. An Optimum GIS Governance Model — Support and Operations
4. Key Performance Indicators (KPI)
5. A High Performance Organization (HPO) — Quality Mindset
6. Reliable and Accurate Digital Data and Databases
7. Training, Education, and Knowledge Transfer
8. Enterprise GIS Software
9. Sustainable IT Infrastructure
10. Mobile GIS
11. Procedures, Workflow, Integration and Interoperability
12. GIS Management (Manager, Director, GIO)
• Selling GIS
• ROI Analysis and Value Proposition
• Consensus Building
• Best Business Practices (BBP)

FIGURE 4.6
A high-performance geospatial organization (HPGO).

6. Reliable and Accurate Digital Data and Databases

7. Training, Education and Knowledge Transfer

8. Enterprise GIS Software

9. Sustainable IT Infrastructure

10. Mobile GIS

11. Procedures, Workflow, Integration and Interoperability

12. GIS Management (Manager, Director, GIO)

 – Selling GIS

 – ROI Analysis and Value Proposition

 – Consensus Building

 – Best Business Practices (BBP)

The seven key characteristics of an HPGO are illustrated in Figure 4.7 are vital to understand. Let's take a look at the following seven key characteristics of an HPGO as they relate to geospatial technology and management in local government.

- Key Factor 1: GIS Organizational Structure and Governance
- Key Factor 2: GIS Teams
- Key Factor 3: GIS Individuals
- Key Factor 4: Geospatial Technology Leaders
- Key Factor 5: A GIS Strategy, Vision, Goals and Objectives
- Key Factor 6: Innovation Geospatial Related Practices
- Key Factor 7: Flexibility and Adaptability

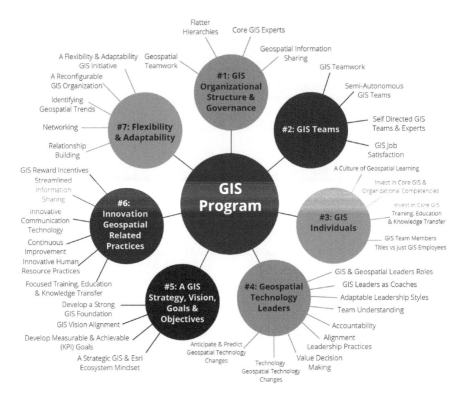

FIGURE 4.7
Seven key characteristics of an HPGO.

Key Factor 1: GIS Organizational Structure and Governance

1. **Geospatial Teamwork**: HPOs value teamwork and collaboration. GIS functional teams for land management, public safety and natural resources are good examples of GIS teams in action.

2. **Flatten Hierarchies**: HPOs flatten organizational hierarchies and make it easier for cross-functional collaboration to occur. GIS and cross-functional collaboration are both critical for success. HPOs remove barriers between functional teams and organizational bureaucracies.

3. **Core GIS Experts**: HPOs introduce experts with distinct functions. This improves organizational performance. GIS management requires specific areas of expertise and lends itself to this type of model.

4. **Information Sharing**: Incentivizing information sharing in both bottom-up and top-down processes is an important characteristics of an HPO.

Key Factor 2: GIS Teams

1. **GIS Teamwork**: The most important difference in an organizational design of HPOs is the reliance on teamwork. GIS programs must enable teamwork.
2. **Semi-Autonomous GIS Teams**: GIS teams operate semi-autonomously to set schedules, manage quality and solve problems.
3. **Self-Directed GIS Teams and Experts**: Self-directed GIS teams are multi-skilled and flexible enough to solve geospatial problems without the need of direct supervision.
4. **GIS Job Satisfaction**: There will be increased job satisfaction with increased self-direction. GIS members of self-directed work teams have greater job satisfaction, more autonomy, input and variety of work.

Key Factor 3: GIS Individuals

1. **A Culture of Learning**: HPOs promote organizational learning, education and knowledge transfer where they invest heavily in their workforce. HPOs typically do this through leadership development and competency management.
2. **Invests in Core GIS and Organizational Competencies**: HPOs develop a clear set of core competencies they want the organization's employees to master.
3. **Invest in Core Training, Education and Knowledge Transfer**: HPOs keep GIS competencies prominent through continued and sustainable training, investment and development.
4. **GIS Team Member Titles versus "GIS Employees"**: HPO organizations reinvent the way they refer to their employees in order to place value on the team concept. GIS team members should be called "team members" as opposed to "employees" or "staff." Again, this simple strategy increases employee involvement and cultivates employees who are committed to the larger goals and competencies that the organization values.

Key Factor 4: Geospatial Technology Leaders

1. **GIS and Geospatial Leaders Roles**: The roles of GIS managers in an HPO should concern themselves with long-term strategic planning and direction. This is a generally hands-off approach.

2. **GIS Leaders as Coaches**: Leaders in HPOs trust in their employees to make the right decisions. Leaders act as coaches to their team members by offering support and directing their employee focus toward the project at hand.

3. Adaptable Leadership Styles: GIS leaders must have the capability to adjust their leadership style based upon the needs of their team members and situational concerns.

4. **Team Understanding**: HPO GIS leaders know when to inspire people with direct communication yet also have the ability to modify management styles.

5. **Accountability**: GIS leaders must hold non-performers accountable reaching their goals.

6. **Alignment Leadership Practices**: HPO GIS leadership practices must be in line with the company's vision, values and goals.

7. **Value Decision-Making**: GIS leaders must make decisions with the organization's values in mind. Leadership behavior must be consistent with the organization's vision, goals and objectives.

8. **Technology Change Decisions**: Leaders in a GIS-centric HPO must understand the changing nature of the geospatial technology world and be able to make bold changes when necessary.

9. **Anticipate and Predict Geospatial Technology Changes**: HPO leaders should have the ability to anticipate changes in the six pillars of GIS sustainability.

Key Factor 5: A GIS Strategy, Vision, Goals and Objectives

1. **Develop a Strong Geospatial Vision**: GIS HPOs require a strong vision, value, mission statement and goals and objectives which guide their local government organizations. Thus vision should be aligned with the organizational and community vision.

2. **The GIS Vision Is a Foundation**: The mission, vision, values and goals and objectives of the local government organization are the foundations upon which GIS initiatives are built.

3. **GIS Vision Alignment**: HPOs implement strategic, tactical, technical, logistical and political vision statements. HPO leaders disseminate this GIS vision to all levels of the organization via knowledge transfer tools.

4. **Measurable and Achievable Goals**: HPOs set the bar high. These goals are measurable and achievable. HPOs use Key Performance Indicators (KPIs).

5. **Strategic Mindset**: The GIS HPO creates a strategic mindset among all team members. Strategy helps an organization achieve its goals.

Key Factor 6: Innovation Geospatial Related Practices

1. **GIS Reward Incentives**: HPOs reward and incentivize behavior that is in line with the organization's goals. They implement reward programs aiming to benefit employees who follow the values of the organization. The values may be expressed by an employee who publishes a paper, presents at a conference, provides exceptional services or completes projects on time and within budget.

2. **Streamlined Information Sharing**: HPOs streamline information sharing across all levels of the organization. There is an open transference of ideas and data sharing. Open exchange is rewarded.

3. **Innovative Communication Technology**: A future GIS HPO should implement innovative Information and Communications Technology (ICT) networks within their local government organization.

4. **Continuous Improvement**: GIS-centric HPOs constantly improve their solutions and services to support local government. HPOs focus on the overall efficiency of their product. HPO use total quality management (TQM) strategies.

5. **Innovative Human Resource Practices**: HPOs have innovative human resources practices. All team members may be involved in the new member hiring process.

6. **Focused Training, Education and Knowledge Transfer**: HPOs focus on the specific training requirements of the organization. Additionally, HPOs have internal training and learning solutions.

Key Factor 7: Flexibility and Adaptability

1. **A Flexible and Adaptable GIS Initiative:** HPOs must be able to rapidly adjust and deploy their existing structures in new environments defined by innovative technology, procedures and protocols.

2. **A Reconfigurable GIS Organization:** HPOs must be able to reconfigure themselves in order to meet the demands of local government and avoid challenges, barriers, pitfalls and organizational threats.

3. **Identifying Geospatial Trends:** A GIS-centric HPO will constantly review and monitor the technological environment to understand new trends and solutions.

4. **Networking:** HPOs promote relationships and partnerships, both internally and externally. HPOs foster close relationships with customers by understanding their customer values and responding to their needs.

5. **Relationship Building:** The maintenance of healthy relationships with identified stakeholders is a core ingredient in an HPO.

5

Value: Business Realization Planning

Any sufficiently advanced technology is equivalent to magic.

– Arthur C. Clarke

Introduction

There is extensive evidence to demonstrate the value, benefits and return on investment when it comes to geographic information system (GIS) and geospatial technology in local government. That being said, it remains difficult for local government organizations to measure and particular monitor the performance of geospatial technology in both quantitative and qualitative terms. Historically, the GIS cost-benefit analysis (CBA), the value proposition (VP) and the return on investment (ROI) analysis have been tools deployed by local governments to evaluate GIS and geospatial technology.

What Is a Cost-Benefit Analysis? "A cost-benefit analysis is a process use to analyze decisions. The process sums the benefits of a situation or action and then subtracts the costs associated with taking that action" (Investopedia, 2019). The outcome quantifies the benefits GIS provides the organization.

What Is a Value Proposition? A value proposition is a pitch to the organization that sells the particular values and benefits of a solution. It aims to convince an organization that a geospatial solution is the best option and offers the most value. It is a declaration of clarity, quality and efficiency.

What Is Return on Investment? According to Investopedia,

> "ROI is a performance measure used to evaluate the efficiency of an investment. ROI tries to directly measure the amount of return on a particular investment, relative to the investment's cost. To calculate ROI, the benefit (or return) of an investment is divided by the cost of the investment."

Typically, a local government identifies the areas where GIS offers the organization quantifiable benefits. For example, the City of Roswell, Georgia, spent significant time and money evaluating GIS improvements to City operations. The following ten examples detail the City of Roswell's success with GIS. Roswell has enjoyed the following successes:

1. Improved information retrieval tools
2. Real-time dynamic databases
3. Innovative thinking and decision making
4. Accurate and reliable record keeping – replaces old record keeping and error reduction
5. New real-time dashboard analytics – big data
6. Automated and advanced field inspections and collection
7. Streamlined processes
8. Navigation and location tools
9. Pattern and predictive analysis
10. Proactive management

The enormous payback successes the City of Roswell experienced can be hard to find in local government. As of August 8, 2019 the City of Roswell secured an Urban and Regional Information Systems Association (URISA) award for exemplary use of GIS in local government. This is a testament to the work performed of Roswell's Mr. Patrick Baber GISP, a GIS coordinator with vision, talent and a commitment to geospatial technologies.

Realizing Geospatial Benefits: An Absolutely Necessary Change in Local Government

An understanding of the business value of geospatial technology and geospatial science's relationship with a local government organization's vision, goals and objectives of a local government organization are all paramount to sustained success.

The GIS team in local government focuses on delivering the right tools and techniques to the right people. Often, they do not focus on the realized benefits. The achievement, accomplishment, realization and attainment of benefits are key factors in the success of a GIS initiative.

What local government doesn't do is benefit planning. Benefit planning must take into account the project's origins, dependencies, alignment with the organizations goals, tangible and intangible benefits, short- and long-term impacts, risks plus timeframe and qualifying metrics for the realization of benefits.

A structured and repeatable strategy for realizing the benefits of geospatial technology is one way to manage the ways programs and projects can add value to the organization. It will make an organization more likely to complete projects on time and within budget. Ginger Levin wrote a conference paper entitled "Benefits – a necessity to deliver business value and culture change, but how do we achieve them?" presented at PMI Global Congress. Levin details the importance of business realization through the lens of definitions and relationship. She explores business realization in a discussion of programs and projects, the first and the ongoing strategy for identification of benefits, the analysis and planning of benefits including key performance indicators (KPI) for all benefits of an organization, and the delivery, transition and sustainment of benefits. Levin also talks about the importance of communicating the necessity of benefits, working with stakeholders, establishing metrics and procedures to define benefits, leadership realization and embracing cultural change. The Levin paper is an exemplary explanation of benefit realization. It includes the "Ten Guidelines for Successful Benefit Realization." Every budding GIO should read and re-read this paper until a commitment is made to enacting cultural change.

In the telecommunications case study at the end of this section, I use the term "FAB." This stands for fact (the problem statement), action (what is the solution) and benefit (the actual quantifiable benefit to the organization). This is a unique way to present the case for business realization.

What Is Benefits Realization Planning?

With the current shift toward smart cites and geo-smart governments, there has never been a more important time to examine the value geospatial science adds to government operations and the community. Business realization planning (BRP) is the next phase in benefit management strategies. It includes a new approach that embraces in-depth analysis of the entire process, supported by some innovative tactics. According to project management informed solutions (PMIS), a BRP "is the **definition, planning, structuring and realization** of the benefits of a business change or business improvement project. It produces the ability to test every single project against the *current strategic objectives of the business* or organization. The central goal of the BRP process is to bring structure, accountability, complete clarity and discipline to the definition and delivery of the benefits of business projects."

This value management philosophy is a fresh strategy for monitoring the entire lifecycle of new business tactics. Customized BRP approaches will pay dividends for local government. I believe that the BRP is a vastly improved strategy and should be the choice of today's new GIO. It encompasses the business case, appropriate measures, benefit drivers and processes and the ongoing monitoring of the benefits of geospatial technology to the organizations. The BRP not only measures how projects and programs add value to organizations, but also provides a set of questions and practices that local government management professionals and leaders can use to help guide the identification, analysis, delivery and sustainment of benefits that align with an organization's strategic goals and objectives.

The PMI thought leadership series defines a BRP by the following strategies:

- Identification of geospatial benefits to determine whether projects or programs can produce the intended business results.
- Execution of geospatial benefits to minimize risks to future benefits and maximize the opportunity to gain additional benefits.
- Sustainability of geospatial benefits to ensure that whatever the project or program produces continues to create value.

BRP is one of the many ways of managing how time and resources are invested into making desirable changes within the organization. A BRP translates analytical results into actionable measures. According to the Association for Project Management (APM) Benefits Management Special Interest Group,

> "If value is to be created and sustained, benefits need to be actively managed through the whole investment lifecycle. From describing and selecting the investment, through program scoping and design, delivery of the program to create the capability and execution of the business changes required to utilize that capability."

Applying BRP to our geospatial world raises four distinct issues:

1. **Organizing and managing projects** so that the benefits arising from the use of geospatial technology are actually realized.
2. The **identification, definition, planning, tracking and realization** of all geospatial technology business benefits.
3. **Initiating, planning, organizing, executing, controlling, transitioning and supporting of change** in the organization so that predefined geospatial technology business benefits are realized.
4. **Implementation of the most valuable initiatives and aligning them with outcomes and business strategies** that increase project success.

"Change management" is an important component of the BRP. This is a significant statement. It's relatively easy to document the benefits of GIS projects. However, it is much harder to monitor and manage the downstream effects and benefits of change.

A large county in the Bay Area of San Francisco identified that a new parcel layer was an essential part of moving forward with a successful GIS. A new parcel layer would offer many benefits to the organization. However, a huge amount of change management was required for such a project. A new parcel layer would have a massive effect on numerous related digital layers, such as zoning and land use.

Today, the normal practice of local government is to simply document the general benefits of a GIS project. What's missing is a structure for realizing the benefits. The following eight components are key building blocks of such a structure:

1. Owners (who are the primary departments?)
2. Dependencies (what inputs are required to make this a success?)
3. Alignment (does this project align with the vision, goals and objectives of the organization?)
4. Benefits (what are the specific quantifiable benefits of a specific project?)
5. Risks (what are the risks associated with a project?)
6. Measuring the benefits (what criteria do we use to quantify and monitor success?)
7. Benefit stakeholders (who benefits?)
8. Benefit tracking (how do we track success?)

Cost-benefit analyses and value proposition assessments all serve a purpose in implementing GIS in local government. However, as we move into the realm of geo-smart government toward an ecosystem defined by the "system of systems," we need an improved and comprehensive way to identify, manage and monitor the benefits of such broad and encompassing technology.

As we approach 2020, we must ask: what are the most common GIS and geospatial projects taking place in local government? Figure 5.1 illustrates a healthy sample of GIS and geospatial projects in local government.

The list detailed in Figure 5.1 illustrates the variety of GIS projects under way in local government today.

Standard Benefit Realization Planning

It was John C. Maxwell who said "good leaders ask great questions." Accordingly, we should consider a few critical standard benefit realization planning strategy questions. Figure 5.2 outlines the following standard benefit realization planning strategy questions:

Common GIS and Geospatial Project in Local Government	
• Create And Refine News Data Layers • Improve Attribution Process • Automated Data Mining And Integration • Refine Active Directory Groups Based On Database Roles • Implement ArcGIS Hub Or Data Collaboration • Redesign Of The Organization Corporate GIS Database • Implement Esri Utility Network Models • Implement Multi-Modal GIS Transportation Model • Data Flow Improvements • 3D Model Of Underground Data And Z-Values • Implement A Geospatial Program • Annual Data Audit • Metadata Initiative • Smart Organization/Real-Time Data Initiative • Smart Structure Monitoring Of Roads, Bridges, And Tunnels • Smart Air Quality Sensors And Heat Island Migration Data And Technology • Establish A Drone Imagery Program • Annual Geospatial Innovation Forum And Report • Master Data List Initiative • Make Governance Changes • Undertake A Project To Clearly Identify Issues With Software Crashes And Make Appropriate Fixes • Create A Mobility Action Plan, Team, And Pilot Project • Master List Of GIS Users And Personal Computer Specifications • Identify And Rectify Speed Issues • Annual Network Testing And Analysis • Fully Implement A Multi Server Isolation Environment • Create A System Design Plan • Implement The System Design • Complete GIS Portal Assessment And Implementation • Full Evaluation And Implementation Of ArcGIS Portal • Holistic Plan And Implementation For A Projects Solution • Plan, Design And Implement Story Maps • Review, Identify, And Prioritize Which ArcGIS Solutions For Utilization • Establish Suite Of Mobile Tools For Field Collection • Implement Esri's Operations Dashboard • Implement Esri's Insights • Migrate To ArcGIS Pro • More Refined Help Desk Flow • ArcGIS For AutoCAD	• 3D Modeling Of Buildings • Contractor Access To Drainage Data • Environmental Construction Waste Pickup Web Application • Evaluate And Adjust Gate Level Enquiry Application • Road Closure And Diversion Applications • Hydrological Analysis • Change Detection • Geo-Enable Social Media • Hire A Geographic Information Officer (GIO) • Adopt A New Hybrid Geospatial Governance Model • Annually Update The Master Business Plan • Adopt Geospatial Vision, Goals, And Objectives • Develop A Coordinated Geospatial Enterprise Program • Create A Geospatial Steering Committee • Formalize A Geospatial User Group • Geospatial Policy And Mandates • Prioritize Geospatial Collaboration • Measure The Quality Of Service • Identify Geospatial Authority And Clear Lines Of Responsibility • Develop An Annual Detailed Geospatial Work Plan • Alignment With An Organization's Corporate Strategy • Annual Update Of The Capability Maturity Assessment (CMA) • Update KPIs Annually • Create A Succession Plan And Strategy • Conduct And Annual Stakeholder Workshop And Review • Standardize Job Classifications • Develop Internal Training Delivery Staff Or Outsource The Delivery Process As Appropriate • Implement An The Organization Specific Training Curriculum To Embrace Existing Training Opportunities And New Opportunities • Implement A Training Communication Plan • Update The Training And Communication Plan On An Annual Basis • Implement Certification Levels Within The Organization For Users Based On A Competency Framework • Sops - Document Each Of The Existing Policies, Procedures, and SOPs, Updated Annually, and Provide Annual Workshops And Training On SOP

FIGURE 5.1
Common GIS and geospatial project in local government.

1. Introductory questions
2. Project dependencies
3. Project alignment
4. Identified project benefits
5. Measuring the benefits
6. Benefit stakeholders

Standard Benefit Realization Planning Strategy Questions	
1. Project Name:	
2. Origin Department:	
3. Benefit Owners:	
4. Project Description / Use Case / Tactical Actions and Operational Processes:	
Project Dependencies	What are the project dependencies?
Project Alignment	Is the project aligned with the organization's Vision, Goals, and Objectives?
Indentified Project Benefits (Explicitly Defined)	a. Tangible b. Intangible c. Short-term d. Long-term e. What are the project benefit risks? Its mitigation?
Measuring Project Benefits	a. When will the benefits be delivered? b. How will the benefits be measured? c. What are the performance metrics or Key Performance Indicators (KPI) of the actual outcome of the project? d. What is the project time frame? e. What is the lifespan of the project benefits? f. How will benefits be reported to stakeholders? g. Are project benefits frequently modified to reflect changing business conditions?
Benefit Stakeholders	a. Have the benefits of the project been clearly communicated to stakeholders? b. Have stakeholders signed off on the benefits? c. Who are the benefit owners? d. Does GIS governance take into account benefits management? e. What is the acceptance criteria?
Project Budget, Funding, and Cost	a. What is the cost of the project? b. Does funding exist for the project? c. Is funding based on the expected benefits?
Benefit Realization, Time Frame / Schedule, and Year	a. Benefit Tracking: Provide tools and techniques to continuously track and report the benefit. (Benefit Planned vs. Benefit Realized) b. Key success factors to achieve and sustain the geospatial benefits c. Map to the Project d. Initiative (related to the gaps): Outcome Benefit Measure e. Benefit Reports and Dashboard

FIGURE 5.2
Standard benefit realization planning strategy questions.

7. Project budget, funding and cost
8. Benefit realization time frame/schedule and year

1. **Introductory questions**
   A. **Project name**: Summary name of each project as identified in previous documents. If we review Figure 5.1, we get an idea of the extent and complexity of local governmental activity.
   B. **Origin department**: The origin department is the department that birthed the idea for a project. Even though it may be IT and GIS making it "happen," the originator of the project is no less important. The Police Department may request a crime analysis

tool or public works may request streaming live video of under-ground pipes. In short, any department can be the originator of the geospatial idea.

   i. **Public Safety and Law Enforcement**: Police, Sheriff, Emergency Management and EOC, Fire Department, Animal Control

   ii. **Public Works and Public Utilities**: Water, Sewer, Storm Water, Solid Waste and Recycling, Engineering, Transportation, Electric, Telecommunications

   iii. **Land and Information Management**: Planning and Zoning Department, Economic Development, Building and Inspections, Code Enforcement, Tax Assessor, Information Technology Department and GIS Division, Public Information Officer (PIO)

   iv. **Natural Resources, Parks and Recreation**: Tree Management and Arborist, Environmental and Conservation, Cooperative Extension

   v. **Public Administration:** Executive Management, Legal Department, Finance Department, Housing Department, Environmental Affairs, Elections

   vi. **Public Services**: Library, Schools, Public Health, Social Services, Community Development

C. **Benefit owners:** Which stakeholders will receive benefit for the project? Who gains from the project? Is it one or multiple departments or is it residents? The benefit owners could include:

   i. **Citizens, public, residents and the community**

   ii. **Commerce, businesses and corporations**

   iii. **Federal, state and local government**

   iv. **Any of the local governments departments**

      – **Public safety and law enforcement**: Police, Sheriff, Emergency Management and EOC, Fire Department, Animal Control

      – **Public works and public utilities**: Water, Sewer, Storm Water, Solid Waste and Recycling, Engineering, Transportation, Electric, Telecommunications

      – **Land and information management**: Planning and Zoning Department, Economic Development, Building and Inspections, Code Enforcement, Tax Assessor, Information Technology Department, Public Information Officer (PIO)

- **Natural resources, parks and recreation**: Tree Management and Arborist, Environmental and Conservation, Cooperative Extension
- **Public administration:** Executive Management, Legal Department, Finance Department, Housing Department, Environmental Affairs, Elections
- **Public services**: Library, Schools, Public Health, Social Services, Community Development

D. **Project description**: A detailed description of each project is important. Descriptions should include a use case, tactical actions and operational processes.

2. **Project dependencies: what are the project dependencies?** Dependencies can take the form of any component of the six pillars of GIS sustainability, including governance, digital data and databases, procedures, workflow, and interoperability, GIS software, training, education and knowledge transfer, and GIS IT infrastructure components. Project dependencies can include the following:

- Data standards
- Data accuracy
- Attribute improvement process
- Training of key staff
- Digital data mining access to windows active registry
- Implementation of Esri's ArcGIS Hub Software
- Real-time access to databases
- Input from stakeholders
- Migration to ArcGIS Pro
- Data reviewer to identify deficiencies in data
- Automatically track record changes
- Documentation of data review procedures
- Validation of data, identification of error trends and correction of data
- Develop metadata templates
- Development of smart technology plan for organization wide implementation
- Systematic deployment of subsequent applications and solutions

3. **Project alignment: is the project aligned with the organization's vision, goals and objectives?** GIS project should align with the overall vision, goals and objectives of an organization. The following are some examples of GIS activities that meet organizational goals.

- **Excellence**: Enabled through more complete and pervasive data sets
- **Accuracy**: Better asset software integration
- **Standards**: SOPs will enable corporation wide standards to be met
- **On-time Services**: Key data viewable in geographic setting assists with ensured on-time delivery
- **Engages Customers**: Stakeholders though GIS software applications
- **Data Sharing**: Shares key information with stakeholders
- **Standards**: Data is standardized and resides in modern geodatabase
- **Enterprise Design**: Database is designed to integrate with existing systems
- **Innovation**: Continued innovation of geospatial programs allow organization to stay current

4. **Identified project benefits (explicitly defined)**

   A. **Tangible benefits**

      - **Benefit: Saving money and avoiding costs** – There is little doubt that GIS results in cost savings and cost avoidance. Immediate savings can be seen through better decision making and increased productivity. Cost avoidance becomes apparent as GIS helps organizations reduce and eliminate duplication of effort.

      - **Benefit: Saving time** – Having the right information when you need it saves time, staff, resources and money. Information can be made available to the public through a web portal, thereby reducing demands on the organization. Geospatial technology offers significant opportunities for time saving.

      - **Benefit: Increased productivity and organizational performance** – Access to accurate, current information, big data analytic tools and predictive analysis will present organization staff with opportunities for increased productivity and performance. Accurate digital data and software solutions will improve accessibility to data. Access and the effective use of GIS tools create high performance and productivity.

      - **Benefit: Improving efficiency** – GIS helps organizations reduce and eliminate redundant steps in workflow processes. GIS programs can reduce workloads and facilitate new procedures that increase productivity and efficiency. The GIS and geospatial ecosystem offers advanced technology to improve existing practices.

- **Benefit: Improving data accuracy and reliability** – GIS creates maps and apps from data. Paper maps can be digitized and translated into GIS. Maps can be created of any location, at any scale and display select information to highlight specific characteristics. Precise GIS data enables users to generate accurate reports and produce quality maps and apps instantly.
- **Benefit: Making better and more informed decisions** – GIS is a critical tool for querying, analyzing and mapping data in decision support. GIS can, for example, be used to identify a location for a development that has minimal environmental impact, is located in a low risk area and is close to a population center.
- **Benefit: Saving lives and mitigating risks** – In an emergency, when every second counts, GIS can support all mission critical services. Organizational data can be used by all stakeholders to make life-saving decisions.
- **Benefit: Automating workflow procedures** – GIS helps automate tasks that expedite workflow and enhance an organization's ability to react efficiently, effectively and responsibly. GIS can automate routine analysis, map and app production, data creation and maintenance, reporting and statistical analysis.
- **Benefit: Improving information processing and analysis** – A true enterprise-wide and scalable GIS streamlines the flow and processing of information throughout the organization. This leads to greater accuracy, increased access, better decision making and increased efficiency in every aspect of the organization. Processing data to make meaningful decisions is of the utmost organizational importance.
- **Benefit: Complying with the vision of the organization** – Digital inventories of an organization's infrastructural assets are becoming increasingly important. A complete GIS program includes asset management, inventory control and depreciation based on accurate and timely data including age, size and construction materials; thus allowing managers to predict and schedule repairs and replacement.
- **Benefit: Protecting the community** – GIS helps officials develop emergency plans and respond to disasters more effectively than ever before. GIS offers tools to monitor conditions, recognize threats, predict consequences and coordinate efficient and effective responses to human-made or natural disasters. GIS can also help officials deliver information to citizens during an emergency via notification systems and the Internet.

- **Benefit: Improving communication, coordination and collaboration between the organization, the community and all stakeholders** – Good communication is the key to running an effective GIS organization. GIS helps staff and stakeholders convey complex information in easy-to-understand formats. The latest geospatial initiatives include extensive public engagement, open and transparent data, and public involvement.
- **Benefit: Provide data to interested parties** – Making digital data available to all stakeholders via the web streamlines and ultimately improves services to the community and organizational departments. Making GIS data available to interested parties will save time, money and resources.
- **Benefit: Respond more quickly to public requests** – GIS data supports intelligent and timely responses to citizen requests information about organization activities, investments and priorities.
- **Benefit: Promotion of an open and transparent government, social engagement and interaction** – GIS-centric Internet solutions, crowdsourcing applications and Esri's advanced new arcGIS hub all streamline citizen access to digital data. Instant access to crucial information is made possible through the use of GIS. New technology allows a two-way interaction with residents.
- **Benefit: Effective management of assets and resources** – Effective management of an organization's infrastructural assets starts with GIS. Tracking, analyzing, managing and conserving assets are key components of asset management.
- **Benefit: Good environmental stewardship and wellbeing** – GIS can promote community wellbeing, healthy populations, environmental protection, community vitality, leisure and cultural education.
- **Benefit: Creating data relationships – new ways of thinking** – GIS offers new ways to think about data. It turns data into meaningful information. New relationships are developing between data, information and decision support. Predicting where and when to replace organizational infrastructure is critical. This is made possible through the improved spatial thinking GIS facilitates.

B. **Intangible benefits**
   - Better insight into issues
   - Staff become conscientious about entering data
   - Enhances reputation of organization

- Assists with sustainability of GIS by implementing standards
- Advances confidence in data
- Provides better and complete data for end users
- Reduces user frustration
- Improves safety of staff
- Cost optimization
- Staff becomes confident in accuracy of data
- Users gain insight into infrastructure
- Easily find data within enterprise GIS
- Connected technological ecosystem
- Improved utilization of enterprise GIS
- Enhance safety performance
- Up-to-date imagery
- Center of geospatial excellence
- Establish accountability for data upkeep
- Increased knowledge allows governance gaps to be filled
- Greater awareness of projects

C. **Short-term benefits**
- Immediate ability to see new data
- Expansion of geospatial insight
- Robust attribution allows better decision making
- Immediate expansion of contributors guided by SOP
- Numerous new data layers available from existing IT systems
- Data security
- Enable customers and stakeholders to "self-serve"
- Standardized data allows for easier data update from field collectors
- Identification of data errors upon load
- Optimal routing based on variables
- Well-defined data flow process
- Shortening of time span from project completion to data assimilation
- Project visualization for data with z-values
- Better decision making based on current data
- Metadata helps keep data accurate and verified
- Metadata provides inventory of data assets
- Digital data produces analytics

– Improved performance of applications
– Improved quality and quantity of collected data
– Improved utilization of desktop GIS solutions

D. **Long-term benefits**
– Optimization of resources based on inventory of existing infrastructure
– Improved data over time by larger workforce contributing
– Improved data due to multiple people correcting errors
– Data security
– Improve rapport with stakeholders
– Improve organizational image
– Better data = better decision making
– New data forced to adhere to business rules
– Various high-quality applications for the public
– Network analysis that incorporates live data
– Better asset management
– Increased confidence in user base
– Increased density of data with z-values
– Ability to analyze how data has changed over time
– Improved data health based on data review/correction
– History of data origin provides documentation of legal issues
– Remote monitoring of assets
– Improved workforce productivity and effectiveness
– Provides foundation for environments to be built on
– Improved system utilization by the public

E. **What are the project benefit risks and its mitigation?** According to Wikipedia,

"A risk–benefit ratio is the ratio of the risk of an action to its potential benefits. Risk–benefit analysis is analysis that seeks to quantify the risk and benefits and hence their ratio. Analyzing a risk can be heavily dependent on the human factor."

The following list was developed by Ms. Reagan Hinton of Geographic Technologies Group. It outlines risks and corresponding mitigation strategies (Figures 5.3 and 5.4).

5. **Measuring the benefits of the project**
A. **When will the benefits be delivered?**
– Immediately upon receipt and acceptance of each data set

- Immediately and continuously as data gets better over time
- Immediately as each IT system is geo enabled
- As user roles are implemented
- As ArcGIS is launched
- As soon as new data model is implemented and data migrated

100 Examples of the Project Benefit RISKS Associated with Not Implementing Geospatial Technology in Local Government	
Risk of continuing to use outdated technology and techniques	Risk that citizens negatively impacted or physically harmed
Risk on not streamline and improve operations	Risk the inability to prevent and respond to emergencies
Risk of duplicating data and data maintenance efforts	Risk of not supporting other key the organization department
Risk of not using geospatial technology appropriately	Risk of not being able to make informed decisions
Risk of failing to understand the benefits of geospatial technology	Risk of not recognizing that the organization can help protect the community
Risk of not avoiding costs	Risk that there will continue to be variations in priorities
Risk of not educating the workforce	Risk that there will be inefficient decision making
Risk of continuing to duplicate work efforts	Risk of poor training and education
Risk of not streamlining and automating manual processes	Risk of insensitivity to user needs
Risk of not having open and transparent government	Risk on everyone doing their own thing and going their own way
Risk of wasting time and resources	Risk of not working as a team
Risk of not improving organizational performance, organizational sustainability, and accountability	Risk associated with damaging the reputation of the organization or undermine confidence in the organization
Risk of not having organization wide resource planning	Risk a lack of strategic decision making
Risk of not using data effectively and efficiently	Risk a lack of stakeholder consent building
Risk of not knowing how to use the latest methods and techniques	Risk a failure of clear lines of responsibilities and accountability
Risk of problems with the organizational culture	Risk of underutilized
Risk of wasting the investment in data and databases	Risk of duplicating effort, time, and resources
Risk of inefficiencies and redundancies	Risk of hoarding data and information
Risk of not avoiding inefficient business processes	Risk on wasting money on data duplication
Risk of not increasing productivity through new technology	Risk of not sharing information to improve operations
Risk of underutilized technology	Risk of not adhering the vision, goals and objectives of the organization
Risk of not applying the right tools to the right problem	Risk of not meeting customer expectations
Risk of stagnation	Risk timely response to citizen requests
Risk of inaccurate information, maps, and statistical reports	Risk of ill-informed citizens
Risk of not automating data capture	Risk of not engaging residents
Risk of poor decision making	Risk of not participating in open and transparent government
Risk of loss of life	Risk of not responding to citizens and residents
Risk of using outdated and costly methods for improving data accuracy	Risk that the public will be negatively impacted
Risk of introducing unreliable geospatial information	Risk to open and transparent democratic government
Risk a lack of strategic decision making	Risk of misinforming the public
Risk of assumption-based decision making	Risk of not seeing GIS information as a public resource
Risk of not turning data into meaningful information	Risk of not reaching out to citizens
Risk of poor decision making	Risk of poor project and process management
Risk of not understanding the relationship of critical data	Risks related to poor decisions about the condition of the organization assets
Risk that community residents will be negatively impacted or physically harmed	Risk of poor economic sustainability and resilience
Risk of not understanding the importance of data and databases	Risk that the physical environment will be damaged
Risk of not using data to mitigate environmental, social or economic impacts	Risk of not promoting the wellbeing of citizens
Risk of not using data to predict infrastructure failure and catastrophic failure	Risk of not promoting healthy populations
Risk of information hoarding and missing information	Risk of not managing infrastructure for parks, leisure, and cultural education
Risk the inability to locate critical or timely information	Risk of not using technology to improve decision support and forecasting
Risk of using outdates and manual processes	Risk of not being able to create tools to analyze data
Risk of not understanding how geospatial technology can automate workflows	Risk of not considering how interoperable CAD, GIS and BIM all work together
Risk of continued data and process duplication	Risk of not promoting the economic opportunities of the organization
Risk of poorly maintained, misplaced, and stale information.	Risk of not using digital GIS data effectively
Risk of not having easy geographic exchange	Risk of not using new geospatial tools to promote local business opportunities
Risk of underutilized data analytics	Risk of not promoting the organization as a high-tech smart government
Risk of not achieving geospatial tasks specifically aimed to the mission, vision, goals, and objectives	Risk of not having the tools to graphically display citizen ideas, thoughts and recommendations
Risk related to the consequences of non-compliance with the vision of the organization	Risk of not exploring how the Organization can make better decision by analyzing data and turning it into meaningful information
Risk of not achieving the stated goals of the organizations geospatial program	Risk of not using technology to be a sustainable & resilient community
Risk of falling short of established goals and objectives	Risk of not using technology to be a more socially engaged community

FIGURE 5.3
100 examples of risks.

What are the project's benefit risks, and what are its mitigations?

According to Wikipedia, "A risk–benefit ratio or a benefit-ration is the ratio of the risk of an action to its potential benefits. Risk–benefit analysis is analysis that seeks to quantify the risk and benefits and hence their ratio. Analyzing a risk can be heavily dependent on the human factor." The following is a list developed by Ms. Raegan Hinton of possible risks followed by their mitigation strategy.

Risk	Mitigation
Incomplete Data Delivered	QA/QC by Data Team
Not Tracking Collection Methods	Design Must Include How Data is Collected, When, and Accuracy
Poorly Designed SOP Leads to Poor Data Improvement	Stakeholders Review SOP and Ensure Staff are Trained
Data Mining App Not Geo-enabling Data	Daily Checks to Ensure Date Stamps are being Refreshed
Unable to Access Key IT Systems	Work Closely with IT to Grant 'Read-Only' Permissions
Poor Geo-enablement from Existing IT — Data is Bad	Create Reject Reports for Data Cleanup, Institute SOP for Data Entry
Inability to Maintain User Roles	Ensure Individual Has This as an Assigned Job Responsibility
Users Put into Wrong Categories	Ensure Help Desk Support and Quick Rectification
Publishing Incorrect Data	Establish Threshold of Completeness and Accuracy
Ignoring Customer Feedback	Install SOP for Rapid Response
Long-Term Fixes Not Adhering to Data Model	Prioritize Data Fixes and Track Progress
Not Having Process for Data Model Update	Ensure Multiple Staff Have the Ability to Augment Data Model
Non-Participation of All Key Agencies	Education Process and Organization-wide SOP
Too Complex/Not Fully Implemented	Detailed Implementation Project Including Tools and Training
Insufficient Z-Values for Key Areas	Implement Z-Value Improvement Program
Management Making Decisions Based on Outdated Data	Manage and Track Changes to GIS Data
Underutilized Data Analytics	Use Data Reviewer Extension to Improve and Configure Data Integrity
Not Maintaining Metadata	Ensure Metadata Process is Audited at Set Intervals
Low to No User Acceptance of Deployed Solutions	Conduct Introduction and Training Sessions
Not Using Technology to Improve Decision Support	Analyze Use of Resultant Data in Predictive Modeling

FIGURE 5.4
Risks and mitigation.

- After acceptance and formalization of new SOP
- Immediately as ArcGIS editor tracking is enabled for each geodatabase feature dataset
- Immediately as each dataset has metadata populated
- As smart technology (sensors) are implemented
- Immediately upon solution deployments
- Over time as more stakeholders buy into solutions
- Immediately upon portal deployments
- Immediately upon problem resolution
- Immediately upon completion of testing and repair/upgrade
- Following implementation of plan steps and recommendations
- Upon completion of elements listed in system design plan
- Annually in geospatial innovation report
- Immediately and continuously as MDL is maintained over time
- Immediately upon successful implementation of project solutions

B. **How will the benefits be measured?**
   - Total number of datasets
   - Total number of feature increase for existing layers
   - Annual data audit
   - Annual Esri data reviewer data audit
   - Quantification of data updated monthly based on time stamps
   - Tracking number of new layers generated in data mining process
   - Tracking number of databases being data mined
   - Yes/no – implemented
   - Tracking number of visits to site
   - Tracking number of partner agencies using the site
   - Full implementation of new data model
   - User satisfaction regarding speed of new model
   - Annual total of participation by key stakeholders
   - Finalization of discovery workshops
   - SOP creation and acceptance
   - Implementation of custom middleware solution
   - Automatically captured statistics generated from data reviewer
   - Metadata plan created and published
   - Analysis of desired end results and benefits
   - Reduction of outstanding issue reports

C. **What are the performance metrics or key performance indicators (KPI) of the actual outcome of the project?**
   - Increase in number of datasets
   - Increase in number of features in datasets
   - Higher percentage of fields and features being populated
   - Number of staff equipped to handle augmentation effort
   - Number of staff contributing
   - High percentage of uptime of data mining tool
   - Total number of layers generated by data mining tool
   - Annual user satisfaction survey
   - Awards won for the site
   - Reduction in data errors as result of noncompliance
   - Number of errors revealed through data reviewer

- Measurement of fully attributed features in database
- Increase in number of key agencies and departments using model
- Tracking decrease in non-delivery of final data from contractors
- Tracking number of users with ArcGIS Pro
- Calculation of increase in z-values within pertinent data sets
- Full implementation of editor tracking
- Fewer errors with each data reviewer run
- Total number of layers reviewed
- Total number of staff trained

D. **What is the project time frame?**

The project timeline can vary from a single week to two years.

E. **What is the lifespan of the project benefits?**

The lifespan of project benefits will vary. It could last:

- In perpetuity
- 2–3 years
- 5-year life cycle for technology
- Five years

F. **How will benefits be reported to stakeholders?**

- Annual data report
- Annual data report highlighting KPIs
- Annual geospatial program report
- Annual metrics report
- Presentations to steering committee
- Workshops
- GIS user groups

G. **Are project benefits frequently modified to reflect changing business conditions?**

- Increase over time as more data sets and features are added
- Modified as each system is geo enabled
- Over time as it systems are cleaned and normalized
- Additional functionality data and apps will be ongoing
- Occasional additions and tweaks to database design to accommodate new data
- Occasional augmentation to SOP to reflect changing conditions

- Increase over time as ArcGIS tracker is enabled for each data set
- Increase over time as metadata entered and standardized
- Technological advances necessitate ongoing modification
- Over time as more feedback is acquired
- Over time as enterprise production environment stabilizes
- Over time as more data sets are added to MDL

6. **Benefit stakeholders**

   A. **Have the benefits of the project been clearly communicated to stakeholders?**
      - Through annual reporting of resolved and unresolved issues
      - Annual solution workshops
      - Feedback should be reported to help determine improvements
      - Annual reporting of downtime events
      - Annual utilization reports
      - Development workshops
      - Users group meetings
      - Annual GIS report
      - Annual meetings to address department needs
      - Annual meetings to discuss ArcGIS pro benefits/usability
      - Regular workshops displaying how users implement both data sources
      - Annual meetings to determine efficiency of project portal
      - Annual report of data created
      - Annual report of usability of application
      - Annual report of number of users
      - Survey of public's response
      - Annual report or maps of change metrics
      - Meetings to determine how geo enabled social media data is being used

   B. **Have stakeholders signed off on the benefits?**
      - Done by client/stakeholder

   C. **Who are the benefit owners?**
      - All departments

   D. **Does GIS governance take into account benefits management?**
      - Operational project

E. **What is the acceptance criteria?**
   - Delivery of completed data layers
   - Data layers with completed set of attributes if discoverable
   - Geodetic team reviews each database prior to committing to production database
   - Review of layers generated from each IT system
   - Review of full inclusion of all users in appropriate Groups
   - Creation of Request for Proposal (RFP) and checklist
   - Review and acceptance of proposed database design
   - Provision of database migration report
   - Review and acceptance of migrated data after testing
   - Review and acceptance of proposed SOPs
   - Review and accept plan to derive z-values from existing plans
   - Review of layers which editor tracking is enabled to ensure being tracked
   - Geodata team review of data review process
   - Sign-off on comprehensive metadata plan
   - Acceptance of smart technology plan

7. **Project budget, funding and cost**
   A. **What is the cost of the project?**

      Multiple projects can cost from as little as $5,000 to multi-million dollar projects.

   B. **Does funding exist for the project?**

      The funding exists in enterprise funds.

   C. **Is funding based on the expected benefits?**

      No

8. **Benefit realization time frame/schedule and year**
   A. **Benefit tracking:** Provide tools and techniques to continuously track and report the benefit. **(Benefit Planned/Benefit Realized)**
   B. **Key success factors to achieve and sustain the geospatial benefits.**
   C. **Map to the project/program**
   D. **Benefit reports and dashboard**

Case Study: Beginning to measure project benefits

GIS benefits for local government telecommunication and fiber to the home (FTTH)

The City of Wilson, North Carolina, owns and operates Greenlight Telecommunication. Even though it is not the norm for local governments to provide telecommunications and fiber to its citizens, the following case study demonstrates the effectiveness of a lightweight and hybridized approach to performing a cost-benefit analysis, value proposition and return on investment.

Let us consider something off the beaten track. As I mentioned, local governments generally don't offer broadband services to citizens. So how does the City of Wilson, NC successfully operate Greenlight Telecommunications? How did they implement this? With what money? And how did GIS help? Well, as a senior staff person at the City told me, "A GIS centric solution gives us real-time monitoring of all our members. We have witnessed improved response time, increased efficiency and effectiveness, and overall a remarkable improvement in customer care, service assurance, and workforce management."

Greenlight uses a remarkable GIS solution

As stated by the Director, Greenlight has "the most advanced telecommunications 'Operational Support Software' in the world. It allows you to launch, sell, bill, manage, and provision voice, video, and data services like never before. It brings together a superior data warehousing solution with Geographic Information System (GIS) technologies."

GIS technology enhances, improves and streamlines all of Greenlight's managerial components, including:

- Customer care and service assurance
- Network planning
- Engineering and construction
- Sales and service delivery
- Planning and market analysis

This is where I created a custom value proposition called FAB, which is an acronym for fact, action and benefit. As the name implies, FAB is a succinct way of articulating the benefits of GIS and creating a value proposition.

Greenlight benefits realization: Customer care, services assurance and workforce management

1. **Reduce churn rate**

   Fact: Subscribers will leave a Service Provider if there are continued calls for service

   Action: Map, in real time, the geography of service calls, repeat service calls and the dashboard statistics of affected subscribers.

   Benefit:
   - Improved speed of issue resolution = saved customers
   - Accelerated service delivery = saved customers
   - Spot patterns and deliver proactive solutions = saved customers
   - Improve response time = saved customers

2. **Improve response time through proactive monitoring**

   Fact: If a Service Provider does not monitor its subscribers and respond in a timely manner, it will lose subscribers and revenue.

   Action: Map and monitor all alarms, work order requests, work orders and work completed using real-time information.

   Benefit:
   - Faster issue resolution
   - Efficient work force mobilization
   - Minimize service down time and service-impacting events
   - Reduce labor and overtime costs

3. **Decrease trouble reports and truck rolls**

   Fact: Hundreds of thousands of dollars will be lost every year if a Service Provider does not optimize field technicians and reduce "truck rolls" or the number of vehicles visiting sites.

   Action: Map the location of field technicians, truck locations and real-time work requests for closest unit dispatch. Map trends and patterns of work order requests and work order completion. Graphically depict data patterns that include specific technical details and chart the incidence of repeat calls.

   Benefit:
   - Workforce management (Training requirement of technicians)
   - Faster issue resolution
   - Lower duration of customer outages
   - Reduction in 'truck rolls"
   - Prioritization of service calls
   - Minimize service down times
   - Improve crew response time and crew training

Greenlight benefits realization: Network planning and management

1. **Predict network demand**

   Fact: If a Service Provider cannot plan, evaluate and predict future demand for telecom services it will not sustain growth or increase revenue.

   Action: Depict existing and future market areas with corresponding demographic data. Map potential homes and businesses that are on the network (close proximity) but are not subscribers. Calculate costs to extend lines.

   Benefit:
   - Target growth areas
   - Selling opportunities
   - Increased revenue
   - Sustained growth

2. **Analyze capacity**

   Fact: The effective use of fiber infrastructure, planned maintenance activities and a real-time understanding of service interruptions will save hundreds of thousands of dollars a year.

   Action: Map fixed telecommunications infrastructure assets and maintenance activities. Map the "utilization" of network elements.

   Benefit:
   - Standardize network asset management
   - Reduce time and costs
   - Improve efficiency of network infrastructure utilization

3. **Boost revenue and improve capital efficiency**

   Fact: Reclaiming fixed assets will minimize operational expenses and save Service providers hundreds of thousands of dollars a year.

   Action: Map fixed assets not being used.

   Benefit:
   - Cost Savings on new infrastructure
   - Savings on inventory warehousing (field inventory)
   - Minimize service down times
   - Improve crew response time

   Fact: The ability to map, visualize and evaluate the entire fiber network will allow Service Provider to design the most efficient architecture possible.

   Action: Geospatially map and query all existing fiber infrastructure.

Benefit:

- Save time by instantly accessing information on logical network expansion
- Increase network revenue by knowing where to expand and gain more customers
- Make informed decisions about the network and its potential return on investment

Greenlight benefits realization: Engineering and construction management

1. **Optimize design and network element management**

   Fact: If a service provider does not keep accurate and reliable fiber records, it will not effectively manage the system and incur hundreds of thousands of dollars in costs.

   Action: Map and manage all network data.

   Benefit:

   - Save time by easily viewing splice diagrams
   - Improve efficiency with direct understanding of alarm impacts on fiber infrastructure
   - Increase network repair efficiency by instantly pin-pointing Optic Time Domain Reflectometer (OTDR) results

Greenlight benefits realization: Sales and service delivery

1. **Analyze service assurance**

   Fact: In order to be profitable, a service provider must be able to analyze current, future and projected revenue while also stimulating service area expansion.

   Action: Map and monitor network members and view viable locations for expansion.

   Benefit:

   - Save time and gain insight by viewing highest average revenue per use (ARPU)
   - Boost revenue by marketing to potential customers in close proximity to existing customers
   - Increase revenue by predicting the most profitable areas to expand network coverage

2. **Discover cross sell, upsell and win back opportunities**

   Fact: A service provider must constantly cross sell, upsell and win back subscribers.

   Action: Map likely candidates for cross selling, upselling and win back opportunities.

   Benefit:
   - Boost revenue by marketing to potential upsell customers
   - Identify and market to past customers to increase revenue
   - Easily identify non-customers with existing infrastructure to promote increased revenue

3. **Study serviceability**

   Fact: A Service Provider must be able to rapidly evaluate the costs and benefits of network expansion in order to systematically plan for growth.

   Action: Quantify the cost of network expansion and view potential return on investments.

   Benefit:
   - Calculate approximate costs of network expansion
   - Save time by easily determining whether a potential customer is within the current network coverage area
   - Prioritize network expansion in the most profitable areas

Greenlight benefits realization: Planning and market analysis

1. **Understanding the market and forecast revenue**

   Fact: A Service Provider must have a "complete revenue profile" of its service area in order to make informed "budget impacting" decisions.

   Action: Overlay existing fiber network with pertinent census information.

   Benefit:
   - Make informed decisions about marketing and network construction
   - Boost revenue by marketing appropriate subscription service levels to appropriate demographic areas
   - Forecast network demand in potential network expansion areas

2. **Manage marketing campaigns**

   Fact: A service provider needs to effectively market its services in order to increase customers and revenue.

Action: Spatially view current and future marketing schemes.

Benefit:

- Easily extrapolate potential customer information
- Minimize marketing and investment mistakes
- Increase revenue by viewing missed homes and businesses

FAB – telecommunications examples

The following summarizes the various benefits of implementing GIS in the management of a telecommunications department:

- **Save time**
  - Automate outage alerts
  - Improve crew response time
  - Minimize service down times
  - Improve information processing
  - Identify past-due and repeat service calls and prioritize field crews
- **Save money**
  - Reduce labor and overtime costs
  - Enhance revenue assurance and optimize income opportunities
  - Reduce churn rate
  - Track capacity levels of each HUB cabinet
  - Effectively reclaim and re-use fiber infrastructure
  - Reduce equipment purchases
  - Field inventory storage
- **Effective management of telecom assets and resources**
  - Standardize network asset management
  - Efficiently tracking and reclaiming infrastructure
  - Improve network infrastructure utilization
  - More informed decision about the network
  - Document integration
  - Improved network repair efficiency
  - Prioritize network expansion
  - Forecast network demand

- **Generate revenue**
  - Track every subscriber on the network
  - Track average revenue per use (ARPU) to show minimum penetration areas
  - Target growth areas
  - Identify new customers
  - Identify logical network expansion
  - Market to potential customers close to, or on the network
  - Upsell to specific customers
  - View gaps in current subscriber locations
- **Improve efficiency**
  - Reduce truck rolls and fuel costs
  - Improve response time
  - Prioritization of service calls
  - Efficient work force mobilization
  - Respond more quickly
  - Faster issue resolution and rapid problem resolution
  - Automate workflow procedures
  - Lower duration of customer outages
- **Improve productivity**
  - Target marketing campaigns
  - Monitor daily work orders
  - Improve communications, coordination, and collaboration
  - Improve dispatching of field technicians
  - Improve management of crews during a storm
  - Accelerate service delivery
- **Make better quality and more effective decisions**
  - Target growth areas and selling opportunities
  - Identifying at-risk customers
  - Better manage customer expectations
  - Track the location of the daily service calls
  - Track all existing customer locations
  - Highlight underperforming areas
  - Identify and resolve issues before a customer is aware
  - Spot patterns and deliver proactive solutions

- **Customer care and service delivery**
  - Churn reduction
  - Proactive monitoring and diagnostics
  - Trouble report and Service Level Agreement (SLA) Management
- **Network planning and management**
  - Demand forecast
  - Capacity analysis
  - Capital efficiency
- Engineering and construction
  - Optimize design
  - Fiber records management
  - Network element management
- **Sales and service delivery**
  - Revenue analysis and assurance
  - Cross-sell, upsell opportunities
  - Serviceability
- **Market analysis and planning**
  - Revenue forecasting
  - Target marketing
  - Campaign management

6

Culture: Sustainability and Resilience

The tools and methodologies are out there. Now they need to be placed into the hands of those who can best use them to assist in creating a more sustainable world for all life on this planet

– Brian Ray James

Introduction

We are becoming ever more environmentally engaged and aware. We will reduce, reuse and recycle ever more resources. We are already making environmentally friendly choices and actively participating in conservation and reducing our carbon footprint by switching off lights, planting trees, composting, conserving water, changing our travel options, using less fossil fuels, buying locally grown products and reducing the use of harmful chemicals. We are protecting our wildlife and our natural habitats. We are trying to educate one another on human activity and the importance of the environment. Sustainability and resilience are parts of our psyche, spirit, soul and consciousness. We are moving in the right direction. Our world is changing for the better. New regulations, new types of investment and social media promote and encourage productive collaboration between government, businesses and the public. Humans already have a greater awareness of our social, economic and environmental future than at any other time in the history of the world.

On a global level, we are shifting toward environmentally sustainable ways of working. The world's biggest companies are disclosing their carbon footprint. Governments are cultivating openness and transparency at all levels. It gets harder and harder to hide illegal and environmentally destructive activities. Oil spills, deforestation and other environmental damage can be monitored in real time using satellite imagery and other technology. Smart mobile phones with high-resolution cameras and video are everywhere. Social media is instantaneous and omnipresent.

Developing better lines of communication and collaboration between scientists, policymakers and the general public is absolutely critical. The future will require every hamlet, village, town, city and country in the world to improve sustainability, increase resilience and understand the full spectrum of technological support required to achieve these goals.

Remember what Michio Kaku (Kaku, 2018) said in his book, *The Physics of the Future – How Science Will Shape Human Destiny and Our Daily Lives by the Year 2100*, namely, that we will rank civilization by its energy efficiency or entropy (or a gradual decline into disorder).

Effective techniques for managing, monitoring and measuring sustainability and resilience will become ever more important. I predict that, by 2025, we will see community sustainability, resilience and smart technology become the primary focus areas of geospatial technology in local government. Towns, cities and counties will measure and monitor success by sustainability variables. Thousands of elected officials are already asking: "How sustainable is our community? And how resilient is our community?" Today the answers to these questions are – for a fleeting moment – focused on solutions related to greenhouse gas emissions, recycling, energy efficiency, renewable energy and waste minimization. In the near future, elected officials will begin asking questions about the budgetary, financial and organizational commitments, required to building out sustainable and resilient communities with long-term strategies for economic growth, social equity, environmental protection and governance models that can handle shifting priorities. There will be a technological shift toward the multifaceted yet unifying solution that underpins the entire concept of smart cities: GIS technology.

GIS and the entire geospatial ecosystem will be used as the primary tools to map, manage, monitor and measure the levels of sustainability and resilience in local government.

So how can we use all of the components of GIS technology in order to evaluate our existing and future sustainability? This is not an easy question to answer, especially considering the ways GIS is currently morphing into a remarkable and pervasive geospatial ecosystem of solutions.

The noble practice of managing sustainability and resilience in local government is still likely seen by elected officials as similar to any other geospatial undertaking. The questions we ask about sustainability and resilience management are uncannily similar to the questions we asked about the business value of geospatial technology in Chapter 5. What is it? How do we measure it? Who benefits the most? And how can we sustain this new approach? As a reminder, the questions we should ask are:

- Name of the project?
- What are the dependencies?
- Is there alignment?

- What are the benefits?
- How will it be measured?
- Who are the stakeholders?
- What is the budget?
- What is the business realization time frame?

There are many other highly qualified practitioners, researchers and authors already documenting, detailing and encouraging open dialogue about the benefits of sustainability and resilience in local government. They discuss the importance of defined outcomes, the meaning of sustainable success, the alignment of an organization's core precepts with sustainability and resilience, the role of stakeholders and community engagement strategies. These dedicated professionals are encouraging many towns, cities and counties to implement sustainable economic development policies and practices that help communities attract and retain jobs while simultaneously improving environmental quality, conserving resources, addressing economic disparities, promoting public health and generally enhancing quality of life for citizens. I will stay away from the many experts who have successfully documented sustainability successes and instead focus on the geospatial science-based aspects of sustainability, resilience and smart technology.

In this chapter, we focus on the future opportunities for GIS and geospatial technology to be deployed as the *solution of choice* for the measurement, tracking, monitoring and reporting of sustainability and resilience in local government. After all, what better set of tools is there to analyze, evaluate and disseminate information in a responsible way? My message in this chapter is simple: GIS and geospatial technology will play a significant role in the level of sustainability and resilience of our communities. Before we proceed, let us consider the following:

- Sustainability in local government
- Resilience in local government
- What are local governments doing to promote sustainability and resilience?
- The future of geospatial technology, sustainability and resilience in local government

Sustainability in Local Government

The term "sustainable communities" is synonymous with the terms "green cities," "livable cities," or "smart cities." Sustainable communities are places where local government meets community needs regarding safety, health and the quality of life. Having said that, today's sustainability strategies tend to be focused on just wind energy, solar power, sustainable construction, green space and sustainable forestry. A select few local governments also train their focus on human, social, economic and environmental sustainability. A "sustainable community" is defined by economic **prosperity, cultural vitality, social equity,** and **environmental awareness and sustainability.**

A common model used to explain sustainability includes the three topics below:

- **Economy**: economic production
- **Society**: well-being and harmony
- **Environmental**: environmental quality

The three components of sustainability are simple yet effective tools for understanding the issues facing local governments interested in sustainability. Researchers say that if any one of the components is compromised, the system as a whole becomes unsustainable. For a community to achieve environmental sustainability, changes need to occur on both societal and economic levels. Some argue that local governments refuse to embrace sustainability as a core strategy because the costs outweigh the benefits. Others say that the trend toward sustainability is becoming – slowly but surely – the main driver for innovation in local government. This change in the mindset of innovators and entrepreneurs will lay the groundwork for our future. In the geospatial world, we've already seen a shift toward social engagement, crowdsourcing, Arc GIS Hub, 3D modeling, improved predictive analysis, story maps, indoor GIS and more.

Resilience in Local Government

A resilient community has the capacity to absorb events, shocks and traumas that strike its social, economic, technical and infrastructure systems while maintaining essentially the same functions, structures and systems. A resilient community manages the risks posed by potential economic crises, health epidemics and uncontrolled urbanization. Take, for example, the opioid crisis currently ravaging the nation. The Centers for Disease Control and Prevention (CDCP) estimates that the total "economic burden" of prescription opioid misuse alone in the United States is $78.5 billion a year. This includes the costs of health care, lost productivity, addiction treatment and criminal justice involvement.

Threats to Local Government

Man Made Disasters: Nuclear, Cyber Attacks, Wars, Riots, Structure Failures (Dams, Oil Spills)

FIGURE 6.1
Threats to local government.

Community resilience is the sustained ability of a local government to utilize available resources to prepare for, respond to, withstand and recover from adverse situations. Communities that are resilient can minimize and mitigate any disaster. The City of Berkeley in California is an excellent example of a city that developed and adopted a resilience plan, despite the fact that developing resiliency plans is not yet a common practice at the local government level. Local government organizations are, however, using geospatial tools to assess their strengths, vulnerabilities and exposure with regards to both natural and manmade threats and then developing strategies and solutions to mitigate impacts. As Figure 6.1 illustrates the various forms threats can take.

The City of Berkeley defined resiliency as "the **ability** of the individuals, institutions, business and systems within the community to **survive, adapt, and grow** no matter what chronic stress or acute shock it experiences. A resilient city lives well in good times and bounces back quickly and strongly from hard times" (City of Berkeley, California 2016).

What Are Local Governments Doing to Promote Sustainability and Resilience?

Very few towns, cities or counties that I have visited over the last 25 years have failed to appreciate the necessity of developing a smart, resilient, sustainable and innovative government and society. Today, towns, cities and counties are thinking evermore about their ability to absorb, recover and prepare for economic, environmental, social or institutional shocks. Towns and cities are promoting a sustainable **economy** with diverse industries and innovative smart tools that support economic growth and innovation.

They are innovating ways to encourage access to employment, education and training. Municipalities are supporting **society** and the **environment** by focusing neighborhood safety and a citizenry kept healthy by investments in parks, recreation and open space. Local government is working hard to meet the challenges of rapid urbanization and growing stress on urban areas.

It is important, however, to remember that the concepts of smart technology, resiliency and sustainability are all relative new phenomenon. That being said, organizations and foundations have been quick to embrace and promote these notions.

- **2016: Rockefeller Foundation 100 Resilient Cities (100RC)** – 100 Resilient Cities (100RC) were selected by the Rockefeller Foundation. A panel of expert judges reviewed over 1,000 applications from prospective cities. "The judges looked for innovative mayors, a recent catalyst for change, a history of building partnerships, and an ability to work with a wide range of stakeholders." It was in 2016 that the final group of cities was announced. One hundred cities now make up this initiative. The 100RC web site describes and details the vision, goals and objectives of this remarkable foundation. Cities in the 100RC network are provided with the resources necessary to develop a roadmap to resilience along four main pathways:
  - Financial and logistical guidance
  - Expert support for development of a robust Resilience Strategy
  - Access to solutions and service providers
  - Membership of a global network of member cities
    http://www.100resilientcities.org/

  We should note that the 100RC organization recently underwent some upheavals and modifications to the organization. Their principle is still sound, and the enthusiasm they fostered for resilient urban areas is here to stay.

- **2016: City of Berkeley, CA, selected as one of the 100RC** – On April 1, 2016, the City of Berkeley released its Resilience Strategy. The Resilience Strategy identifies six long-term goals and recommendations for specific short-term actions to help address some of Berkeley's most pressing challenges. https://www.cityofberkeley.inf o/Resilience/

- **2017: World Bank City Resilience Program (CRP)** – Established in June 2017, the World Bank Group's CRP empowers cities to pursue investments that build greater resilience to climate and disaster risks and provides access to the financing necessary to ensure that those investments come to fruition. https://www.worldbank.org/ en/topic/disasterriskmanagement/brief/city-resilience-program

The Rockefeller Foundation 100 Resilient Cities (100RC) and the World Bank: CRP offer glimpses of the future focus of local government. Today, local government organizations and authorities are just begging to understand and implement resiliency and sustainability programs.

Hazard mitigation that includes preparedness for, response to and recovery from natural and human-made disasters has been around for a long time. This is a close and important cousin of sustainability and resilience. The coming era will require resourcefulness, creativity and innovation. It may include new planning policies, new regional growth strategies and a shift in the way local government professionals support economic, social and environmental initiatives. Figure 6.2 outlines a few of the sustainability and resilience challenges faced by local government:

Sustainability and Resilience Challenges Faced by Local Government
Balance economic, social, cultural and environmental interests
Capacity to adapt and change to shifting events, incidents, occurrences, and aftermaths
Ability to minimize waste and inefficiency
Plan to reduce future costs of infrastructure
Support community safety, health and social equity and diversity
Develop a blueprint for a community that promotes physical activity, community connection, affordable housing, security, substance abuse reduction, cultural diversity and accessibility to services
Support the natural environment and its ecosystems
Reduce greenhouse gas emissions, protect agricultural land and green space, and maintain wildlife corridors, habitat and values.
Control urban sprawl
Promote mixed land use that supports the efficient movement of people, goods and services and contributes to business efficiency and quality of life.
Design focal points and population center within a city that support business and gatherings
Propose linked development that includes transport routes and public spaces
Create Socially Connected City
Enforce Community Safety City
Environment as a natural asset, ensuring a productive resource base that includes agricultural and forest land, and protecting assets such as tree canopies, streams, groundwater and aquifers
Mitigate natural and man-made hazards: Planned development that will reduces vulnerability to hazards including fires, hurricanes, blizzards, more frequent intense storms, and sea level rise.
Promote Residential and Commercial Recycling: Recycling keeps recyclable material out of landfills. This conserves resources and extends the landfill's life. It saves businesses money and reduces the potential generation of methane, a greenhouse gas that is released from improperly managed landfills.
Create Healthy Communities: Farmers markets, pedestrian and bicycle friendly urban areas, a balance of housing, jobs, shopping, schools and recreation give are all part of well-planned communities.
Encourage Green Building: Green building policies and ordinances reduce energy consumption.
Promote Local Green Businesses: The promotion of local green businesses also allow a local government to conserve energy and water and minimize waste and prevent pollution.

FIGURE 6.2
Sustainability and resilience challenges faced by local government.

- **Balance** economic, social, cultural and environmental interests
- **Capacity** to adapt and change to shifting events, incidents, occurrences, and aftermaths
- **Ability** to minimize waste and inefficiency
- **Plan** to reduce future costs of infrastructure
- **Support** community safety, health and social equity and diversity
- **Develop a blueprint** for a community that promotes physical activity, community connection, affordable housing, security, substance abuse reduction, cultural diversity and accessibility to services
- **Support** the natural environment and its ecosystems
- **Reduce** greenhouse gas emissions, protect agricultural land and green space, and maintain wildlife corridors, habitat and values
- **Control** urban sprawl
- **Promote** mixed land use that supports the efficient movement of people, goods and services and contributes to business efficiency and quality of life.
- **Design** focal points and population center within a city that support business and gatherings
- **Propose** linked development that includes transport routes and public spaces
- **Create** socially connected city
- **Enforce** Community safety city
- **Environment**: Ensuring a productive resource base that includes agricultural and forest land while protecting assets such as tree canopies, streams, groundwater and aquifers
- **Mitigate natural and human-made hazards**: Planned developments that reduce vulnerability to hazards including fires, hurricanes, blizzards, more frequent and intense storms and sea level rise.
- **Promote Residential and Commercial Recycling**: Recycling keeps recyclable material out of landfills. This conserves resources and extends the landfill's life. It saves businesses money and reduces the potential generation of methane, a greenhouse gas that is released from improperly managed landfills.
- **Create Healthy Communities**: Farmers markets, pedestrian and bicycle friendly urban areas, a balance of housing, jobs, shopping, schools and recreation give are all part of well-planned communities.
- **Encourage Green Building**: Green building policies and ordinances reduce energy consumption.
- **Promote Local Green Businesses**: The promotion of local green businesses allows a local government to conserve energy and water, minimize waste and prevent pollution.

Local governments are starting to understand the economic benefits of sustainability. Towns, cities and counties are investing in energy efficiency and alternative energy while planning and designing vibrant, mixed-use downtowns. We must, however, be realistic about the technology and functionality of today's geospatial ecosystem. We do not have a standardized strategy or methodology for the practical use of desktop, web, and mobile geospatial technology to measure and monitoring sustainability and resilience.

The Future of Geospatial Technology, Sustainability and Resilience in Local Government

If the goal of local government is to reach optimum sustainability and resilience, then the multifaceted geospatial tools for collecting, assembling, grouping, organizing, analyzing and disseminating the massive amounts of social, economic and environmental data necessary for evidence-based decision making are the most important tools in our shed. Also – and this is a big also – we must build easy-to-use tools that make it easy for all interested parties to understand sustainability and resilience in our towns, cities, counties, regions and states. If we managed to build a time machine but no one used it, what has really been accomplished?

On a more positive note, in many places this process is already under way. In fact, the catalyst for my own interest in the sustainability, resilience and livability of cities originated between Canada, California and Indiana. I was working for two Canadian cities (the City of Mississauga and the City of Nanaimo) and two American cities (the City of Berkeley, California and the City of Hobart, Indiana). The Canadian cities were focused on sustainability while the City of Berkeley was focused on resilience planning. The City of Hobart had just completed an environmental sub-plan that focused on sustainable neighborhoods. Ironically, even though the City of Hobart is not coastal, this study was funded by a grant from the National Oceanic and Atmospheric Administration and the Indiana Department of Natural Resources Lake Michigan Coastal Program. If these are harbingers of things to come, we need to put our geospatial ideas, tools and methodologies that have been developed through academic, private and government-funded research in the hands of the policymakers, planners, GIS specialists and engineers who can present the outcomes of an analysis.

Let's ask a few questions. How is geospatial science and technology *relevant* to sustainability and resilience in local government? How is geospatial *thinking* – including spatial distributions, spatial interactions, spatial relationships, spatial comparisons and temporal (time-based or historical or chronological) relationships incorporated into sustainable and resilient local governments? More importantly, where do we describe the world of geospatial thinking from the perspective of local government?

Considering, measuring and monitoring the sustainability and resilience of a community marks a cultural shift. It offers a new way to think about managing our assets, people, resources and vulnerabilities. It's a key component of a smart city. It is something that must be embraced sooner rather than later.

We can begin by using the six pillars of GIS sustainability (Chapter 1, Figure 1.15) to guide our conversation. We must change our style of GIS governance, our analysis of digital data and databases, a re-thinking of existing procedures, workflow and database integration and interoperability. We need a renewed strategy for using the entire GIS and Esri software ecosystem, supported by a strategy for training and knowledge transfer throughout the organization coupled with a movement toward real-time systems and sophisticated IT Infrastructure.

GIS and geospatial science will guide towns, cities and counties down the road to sustainability and resilience. A new framework for evaluation will help identify the areas most in need of sustainability measures. The Multiple Criteria Analysis (MCA) of social, economic and environmental must become more original, creative and innovative. A defining characteristic of a smart community will be its measurable levels of sustainability, resilience and livability index. The following are recommended changes to the existing GIS environment that will prove necessary if we aim to take sustainability and resilience planning seriously.

- **A New Governance Mindset**: The GIO has a new set of tasks to think about in the near future, including the management and maintenance of data standards for sustainability and resilience, the creation of new data layers and databases and the regionalization of a governance model that specifically focuses on needs of sustainability and resilience. The GIO will have to organize the interoperability of systems, the creation of new types of predictive analysis, the building of software applications and the delivering of an ecosystem of solutions to communicate sustainability and resilience findings and recommendations. Additionally, they will organize the training, educating and knowledge transfer necessary for sustainability and resilience.

- **Multiple Criteria Data and Database Analysis (MCA)**: The amount of real-time and temporal data *related to the three main areas by which to measure sustainability (economy, society and environment)* is vast and dynamic. The sheer amount of data, coupled with the MCA and the modeling and programming of algorithms necessary to support sustainability and resilience planning, will require changes in local government. The situation will necessitate new "derived" data sets, analytical routines, real-time feeds.

- **Systematic Framework of Procedures, Workflow, Integration and Interoperability**: Sustainability and resilience planning requires new procedures for data quality and content. The IT infrastructure requires a new workflow for collaboration between agencies and custodians of data. The city of tomorrow's IT infrastructure will require a smart city approach to interoperability that allows access to all databases within and beyond an organization. The abundance of data comes from many different software solutions and sources. These include the federal government, state government, and local and private organizations. It is a truly collaborative interoperable strategy.

- **New Predictive and Real-Time Decision Support and New GIS-Centric Software Applications**: A combination of desktop, web and mobile solutions, in addition to real-time data feeds, are required to change the ways we evaluate the sustainability and resilience of our towns, cities and counties. Web applications and dashboard solutions to monitor our sustainability index will be part of our future. Sustainability, resilience and a livability index will not be confined to a specific geography. Any user-defined geography should be easily measured for sustainability, resilience and livability. We will see three types of applications:
  - Sustainability Application
  - Resilience Application
  - Livability Index Application
  - Each will be supported by Esri's total GIS ecosystem used for collecting, analyzing and disseminating information to the public.
- **New Style of Training, Education and Knowledge Transfer**: The procedures, methodology and scientific approach to measuring sustainability and resilience of our local government will require new strategies for training, education and knowledge transfer. The myriad GIS software tools, not to mention the idea that we can monitor sustainability and resilience in real time will be foreign to many professionals. This necessitates an overhaul of existing training and education practices in local government. It will require uniform and accepted GIS practices.
- **New Real-Time, Remote Sensor, System of System IT Infrastructure**: The variety of digital data required to monitor the sustainability and resilience of a local government is abundant. This smorgasbord includes data about air quality, the condition of aging infrastructure, the types of land use, the incidence of crime and more. The Green City Index series uses thirty indicators including CO_2 emissions, energy, buildings, land use, transport, water and sanitation, waste management, air quality and environmental governance to measure sustainability (Figure 6.3).

The future of cities lies in measuring and monitoring sustainability!

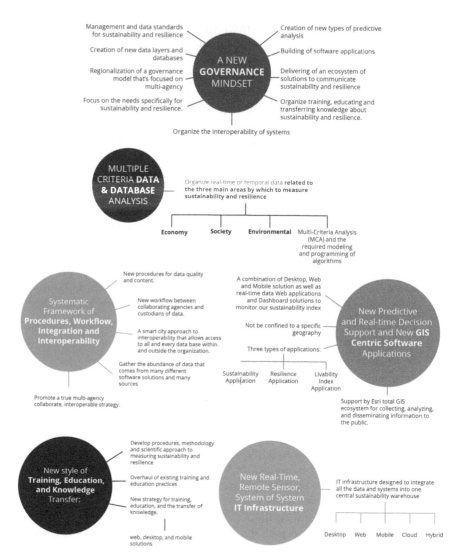

FIGURE 6.3
Sustainability and resilience planning needs to change the geospatial landscape.

A Case Study of the City of Hobart Sustainable Neighborhood Plan: A Move in the Right Direction

I developed a particular interest in The Hobart Sustainable Neighborhoods: Environmental/Ecological Sub plan. It stated that the three factors integral for sustainability are ecology, economy and equity. This plan focused on the ecological portion of sustainability. The five main components of this GIS-centric sustainability plan included:

1. What do we mean by sustainability?
2. **The objective of the Hobart Sustainable Neighborhoods (HSN) plan**: The objective was clear. It was to "advance sustainability one neighborhood at a time by identifying and benchmarking each residential area to pinpoint neighborhoods suffering from unsustainable practices."
3. The GIS Process
    * Acquiring Data
    * Preparing Data
    * Organizing Hobart into Tracts, Groups and Neighborhoods
    * Benchmark Hobart's Environmental Sustainability
    * Conducting Sustainability Analysis and Identifying Problem Areas
4. The Geospatial Benchmarks for Environmental Sustainability Analysis and Solutions
    A. Environmental Hazards
        - Clean-up sites
        - Institutional control sites
        - Underground storage tanks
    B. Sanitary Sewer lines, Facilities and Water Wells
        - Septic Tanks
    C. Storm Water Lines and Facilities
    D. Impervious Surfaces
    E. Land Cover and Open Space
    F. Managed Lands, Parks, Trails and Recreation Facilities
    G. Waterways and Impaired Water Ways
    H. Tree Canopy
    I. And many more
5. Offering of Solutions

The success of the Hobart project lies in the fact that the process of acquiring data, geospatially analyzing data, identifying and isolating areas of concern and the offering solutions from sources like the EPA Guide For Local Decision makers titled "Getting to Green: Paying for Green Infrastructure" (2014) is a repeatable and sensible methodology. The Hobart plan explained that the solutions for areas identified in the study as non-sustainable were drawn from four distinct sustainability guidebooks and various credible Internet resources.

Evidence confirms that we are witnessing unprecedented migration of people to cities. In some areas of the world, the uncontrollable urban sprawl has already introduced air quality problems, climate change issues, water pollution, increased crime rates, traffic congestion, public health concerns and economic challenges. There is real fear that this situation will become the norm.

We have an urgent need to standardize our methods for collecting, storing, managing and analyzing vital data that accords with industry and government-endorsed models.

It is worth looking at the Green City Index. The Green City Index methodology was developed by the Economist Intelligence Unit (EIU) in cooperation with Siemens.

This is the start of things to come. The GIS and geospatial community needs to stand up and develop standard and government accepted tools to calculate sustainability and resilience.

7

Applications in Local Government: A Smart Geospatial Ecosystem

Mystery creates wonder, and wonder is the basis for man's desire to understand. Who knows what mysteries will be solved in our lifetime, and what new riddles will become the challenges of the new generations?

– Neil Armstrong, 1969

Introduction

Ask a local government geographic information system (GIS) manager what aspect of working with GIS is the most fun. They will almost surely answer: software applications. When a GIS software solution automates a traditionally manual process or computes something that would traditionally take a million hours to calculate, GIS managers beam with pride. No person is immune to the joy of making a difference! No person dislikes solving a mystery!

Chapter 7 examines today's GIS software ecosystem and the role of smart software applications in local government. First, we should discuss today's GIS software solutions in terms of desktop, web, mobile and enterprise architecture tools. Second, it is important to understand Esri's ubiquitous and extensive GIS ecosystem through the eyes of the user. Third, Chapter 7 documents twenty-one of the most popular GIS applications used in local government today. The final section of Chapter 7 is a Smart Park case study documenting how an enterprise "bespoke" GIS solution can change the way Parks, Recreation, Open Space and Natural Resource managers think about geospatial technology.

Today's GIS Software Solutions Desktop, Web, Mobile and Enterprise Architecture Tools

Environmental Systems Research Institute (Esri) was founded in 1969 and is – without a doubt – the world's leading supplier of GIS software and solutions. Headquartered in Redlands, California, the growth of Esri remains remarkable. Today, Esri is said to be a $1.1 billion company with nearly 4,000 employees. The complete Esri ecosystem of geospatial solutions is complex and multifaceted. Even Esri employees find it hard to compartmentalize the vast amount of products and solutions on offer. If we, just for a moment, temporarily ignore all the enterprise architecture tools the Esri Corporation offers, we are left with three specific application areas including:

- Desktop Solutions
- Web Solutions
- Mobile Solutions

Within each system of "geospatial delivery," we find an abundance of add-on tools. The following is a description the GIS world as seen by the end GIS user. A number of professionals could certainly argue that the GIS ecosystem far more interconnected and complex than my interpretation. I will on this point, however, refer to a Thomas Mann quote: "Order and simplification are the first steps toward the mastery of a subject." In the interest of "simplifying the complicated," we need to talk about Esri GIS software in terms of desktop, web, mobile and other enterprise architecture solutions. Figure 7.1 diagrammatically illustrates my user-eye-view of the Esri GIS software ecosystem. It focuses on compartmentalized solutions.

Esri's Ubiquitous and Extensive Ecosystem

Esri has produced a total suite of GIS software solutions that meet virtually every GIS requirement of local government. It includes sophisticated desktop functionality, web-based architecture that enables a wide population to

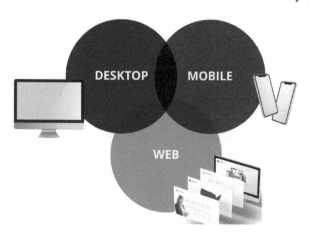

FIGURE 7.1
A simplified Esri GIS software ecosystem.

interact with data, and mobile solutions that offer field-to-office and office-to-field capabilities. Below, I've detailed the Esri suite of solutions and the tools that make it work.

Desktop environment includes GIS software solutions, in particular ArcGIS Pro, business extensions and intelligence, data conversion tools, drone interfaces, spatial analysis routines, data and system interoperability solutions and indoor mapping.

Web-based GIS solutions include custom web application building for internal business operations and external public engagement and crowd-sourcing consumption, ArcGIS online, operation dashboards, advanced Open Data (OD) HUB solutions, Story Mapping, Insights for data analysis and interpretation, Business Analyst and Community Analyst, Dynamic Integration and Interoperability, Data and Spatial Analytics Using Web Tools, Geo-Planner, Survey123 Web, ArcGIS Photo Survey and ArcGIS Urban.

Mobile solutions include ArcGIS Online, Collector for ArcGIS, Survery123, Operations Dashboard, Business extensions for Workforce, Navigator, general Explorer, and Dynamic Integration and Interoperability, Data and Spatial Analytics, and ArcGIS Companion and Trek2There.

Enterprise architecture tools include ArcGIS Enterprise (Web GIS Deployment) and ArcGIS Data Store, ArcGIS Server, ArcGIS Web Adaptor and Portal for ArcGIS. There are also Data Catalog, Secure Departmental Data Sharing, Secure Mobile Apps, Data and Spatial Analysis, Dynamic Integration and Interoperability, Relational Database Management (RDB), Real-Time Data Management, including Geo Event Extension, AVL (Automatic Vehicle Location) Services and the Internet of Things (IoT) and Sensor Services. Additionally, there are ArcGIS Monitor, ArcGIS License Manager and Application Development that include the Web App Builder for ArcGIS, ArcGIS App Studio, ArcGIS APIs, ArcGIS Runtimes and Libraries & Extensions.

Figure 7.2 introduces a larger Esri GIS ecosystem of solutions.

Smart Local Government Software Applications

There are two types of software initiatives in local government:

1. Enterprise Software Solutions: used by and for the entire organization
2. Departmental Software Solutions: used by specific departments

Let's discuss both these initiatives in terms of the Esri software suite.

GEOSPATIAL SOFTWARE ECOSYSTEM

Enterprise Architecture | Desktop | Web | Mobile

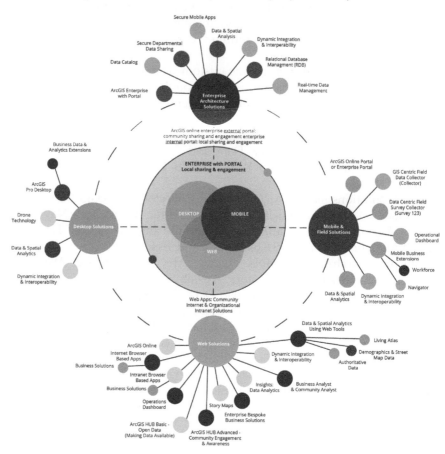

FIGURE 7.2
A larger Esri GIS ecosystem.

Enterprise Applications

Enterprise GIS applications meet the broad needs of individual departments, the organization as a whole and the population served by local government.

- **My Government Services (Figure 7.3)**
- **Parcel Notification (Figure 7.4)**

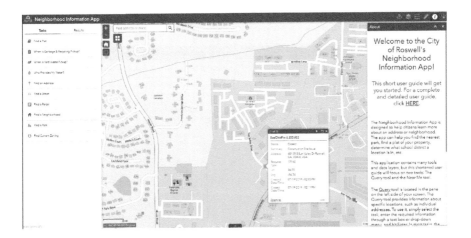

My Curbside Services: City of West Hollywood, LA

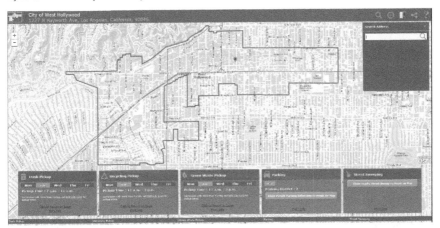

FIGURE 7.3
My Government Services: City of West Hollywood, CA.

- **Arc GIS HUB – Open Data Sharing and Transparent Government (Figure 7.5)**
- **Crowdsourcing Applications (Figure 7.6)**

Departmental Applications

Public Safety and Law Enforcement: Police, Sheriff, Emergency Management and EOC, Fire Department, Animal Control

FIGURE 7.4
Parcel notification: City of Berkeley, CA.

FIGURE 7.5
Open data sharing: City of Roswell, GA.

Public Safety is one local government's most significant and complex functions. It encompasses a wide range of services. Esri-based GIS and analytical software solutions are used by Police and Sheriff Departments, Fire and EMS, Communication Centers and Emergency Operations Centers (EOCs). GIS technology is used to display patterns of crime and fires, predict where crime will occur, view what Tweets, display real-time weather feeds, perform

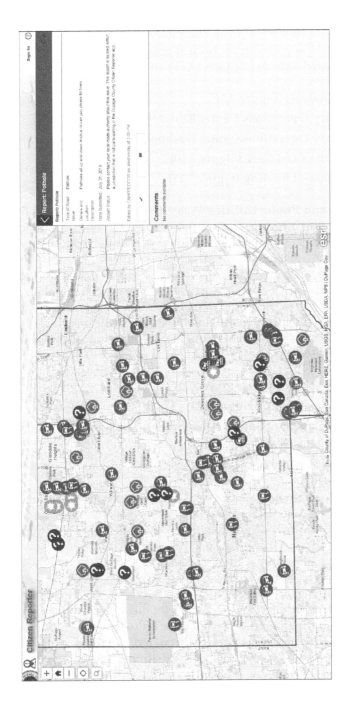

FIGURE 7.6
Esri crowdsourcing application.

database analysis, conduct citizen outreach and social engagement projects, create dashboards and up-to-date and real-time statistics, view a Common Operating Picture (COP) of all public safety data, deploy story maps, utilize the Collector application for hazard mitigation, use dispatch software, deploy Automated Vehicle Location (AVL) and more.

- **Crime Analysis (Figure 7.7)**
- **Emergency Operations Center Browser (Figure 7.8)**
- **Emergency Operations Center Dashboards (Figure 7.9)**
- **Story Map –Cold Case (Figure 7.10)**
- Story Maps – Flooding (Figure 7.11)

Public Works and Public Utilities: Water, Sewer, Storm Water, Solid Waste and Recycling, Engineering, Transportation, Electric, Telecommunications

Public works departments offer many services to the community including the planning, design and construction of public buildings, transportation infrastructure (roads, railroads, bridges, pipelines, canals, ports and airports), public spaces (public squares, parks, beaches), public services (water supply and treatment, and dams) and other long-term physical assets and facilities. Public utility maintains the infrastructure for a services consumed by the public that include but are not limited to electric, gas, water, sewer, telephone, transportation and broadband internet services.

- **Utilities Web Browser (Figure 7.12)**
- **Utilities Dashboard (Figure 7.13)**
- **Fiber Management Solution (FMS) (Figure 7.14)**
- **Fiber Analysis Web Browser (Figure 7.15)**
- **Work Force Solution (Figure 7.16)**

Land and Information Management: Planning and Zoning Department, Economic Development, Building and Inspections, Code Enforcement, Tax Assessor, Information Technology Department, Public Information Officer (PIO).

The land and information management category encompasses the legal enforcement of land use and land-related development regulations. It includes planning, land use and zoning activities, the promotion and retention of businesses through economic development, permitting and regulating through building inspections and code enforcement. Additionally, this category includes parcel management through the Tax Assessor, technology support from Information Technology Department, and the promotion and communication through the Public Information Officer (PIO).

- **Economic Web Application (Figure 7.17)**
- **Planning and Zoning – 3D Urban Solutions (Figure 7.18)**

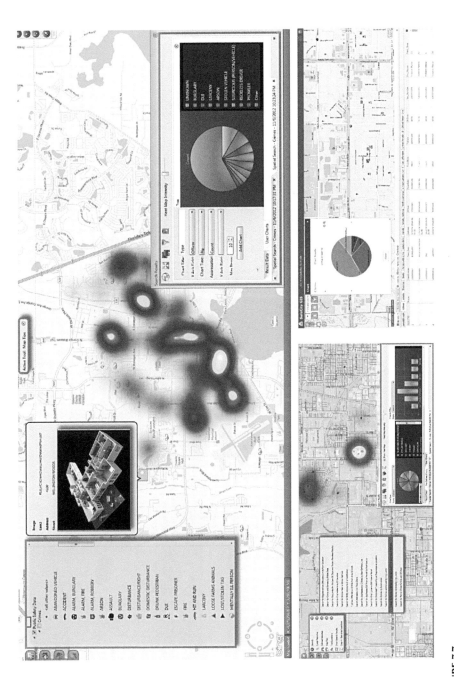

FIGURE 7.7
Crime analysis: City of Roswell, GA.

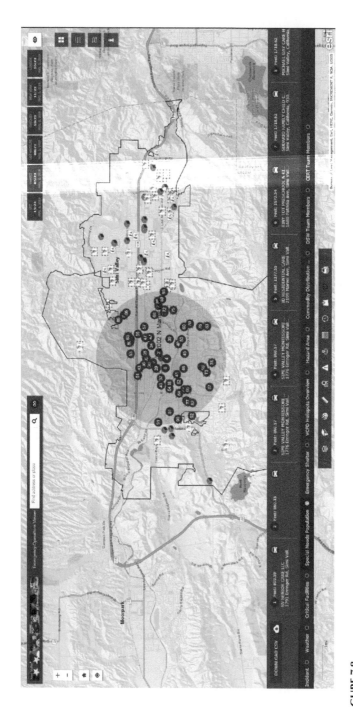

FIGURE 7.8
Emergency Operations Center.

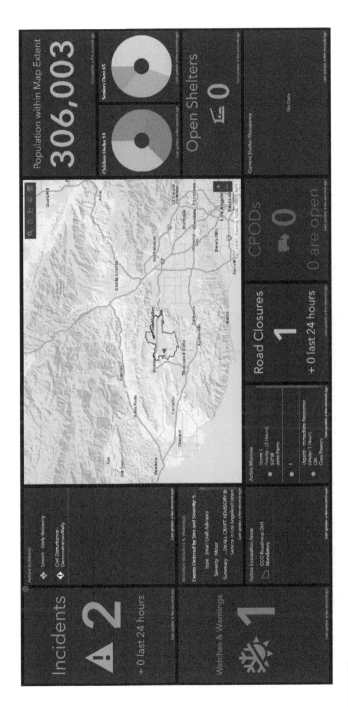

FIGURE 7.9
Emergency Operations Center dashboard.

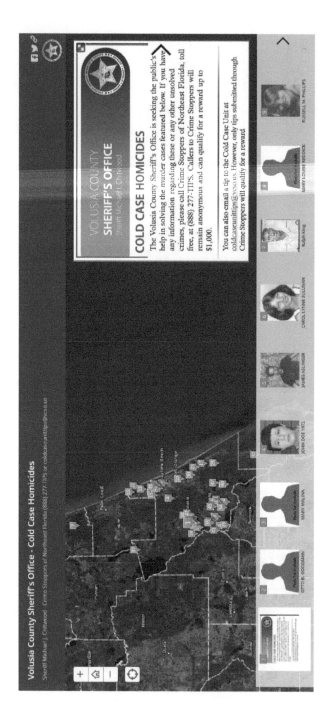

FIGURE 7.10
Cold Case Story Map – Volusia County, FL.

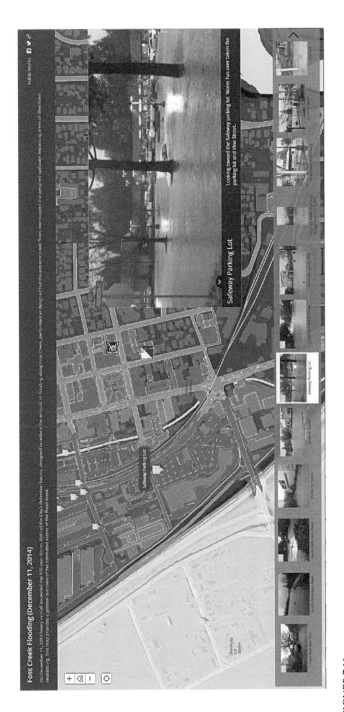

FIGURE 7.11
Flooding Story Map: City of Healdsburg, CA.

FIGURE 7.12
Utilities Web Browser – Town of Windsor, CA.

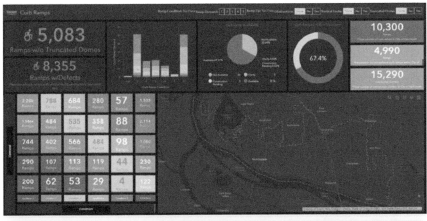

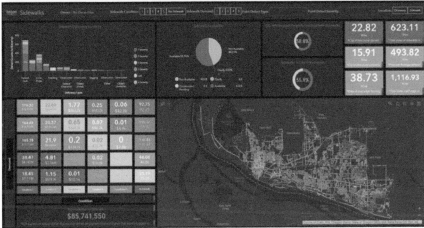

FIGURE 7.13
Utilities Sidewalk Dashboard – City of Vancouver, WA.

A Case Study of Natural Resources, Parks and Recreation: Tree Management and Arborist, Environmental and Conservation, Cooperative Extension

The Smart Park case study documents how an enterprise 'bespoke" GIS solution can change the relationship between Parks, Recreation, Open Space and Natural Resource managers and geospatial technology.

Case Study Introduction: This following case study describes GIS software applications used in parks and recreation departments to support improved sustainability and resilience, high-performance organizations (HPO), business realization planning (BRP) and smart parks.

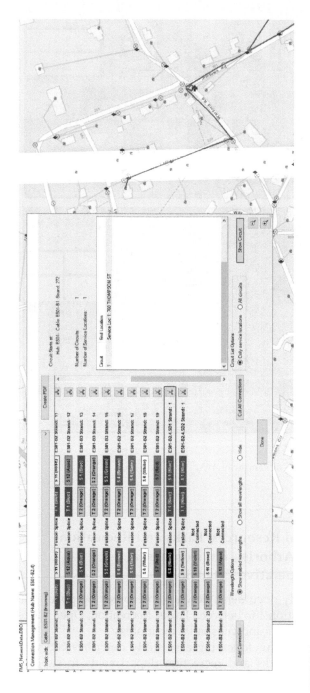

FIGURE 7.14
Desktop FMS – City of Wilson, NC.

FIGURE 7.15
Fiber Analysis Web Browser: City of Wilson, NC.

FIGURE 7.16
Work Force: City of Berkeley, CA.

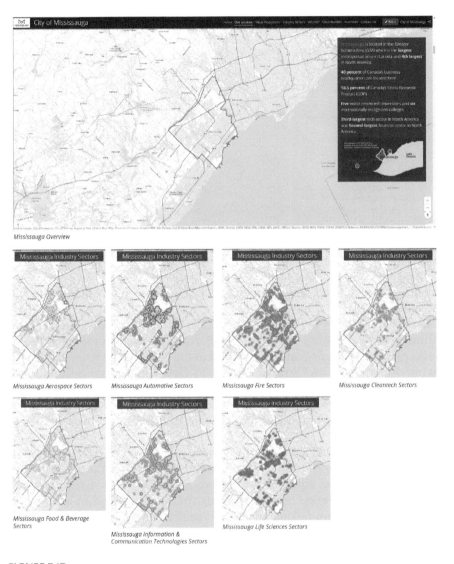

Mississauga Overview

Mississauga Aerospace Sectors   Mississauga Automotive Sectors   Mississauga Fire Sectors   Mississauga Cleantech Sectors

Mississauga Food & Beverage Sectors   Mississauga Information & Communication Technologies Sectors   Mississauga Life Sciences Sectors

FIGURE 7.17
Economic Web Browser: City of Mississauga, Ontario, Canada.

During the writing of this case study, the Mayor of London, Sadiq Khan, announced the next step toward making London the world's first National Park City. His plans include a week-long National Park City Festival to celebrate London's **green spaces**, **wildlife**, **green rooftops** and **waterways** and help a wide range of Londoners discover new **environmental**, **cultural**, **sporting** and **community activities**.

Park History: In 1634, the Boston Common was designated as the first United States open space city park. According to the history books, sixteen

FIGURE 7.18
3D Urban: City of Roswell, GA.

other city parks were created before 1800, including the National Mall in Washington, DC.

Early Parks: Parks in the 1800s centered on the urban park vision of providing a natural setting for urban dwellers. There were some magnificent parks planned, designed and constructed in the 1800s by the noted landscape architect Frederick Law Olmsted. Olmsted's view was that parks should provide a natural, pastoral environment where city residents could escape the bustle of urban life. Frederick Olmsted parks included Central Park in New York City, Washington and Jackson Parks in Chicago, and Prospect Park in Brooklyn.

The Neighborhood Park: Beginning in the early 1900s, the vision for urban parks began changing. There was a demand for parks that provided recreation and playground equipment for children. Parks of this style are known as "neighborhood parks." Eventually, this concept expanded to include the swimming pools, ball fields and indoor facilities that cropped up in parks during the early and middle parts of the twentieth century.

Migration to the Suburbs: The post-war period saw the urban population move to the suburbs. After 1945, the public interest in city parks waned. In this period, people wanted a yard of their own. This made the notion of public parks nearly obsolete. Many urban city parks declined well into the 1990s.

City Park Revival: City parks did not experience a revival until the 1990s, when the populations of urban areas began again to grow. Parklands, recreation, open space, critical habitat and natural resources within our communities have become more important over the last twenty years. The 1990s also witnessed a corresponding and significant uptick of GIS technology in local government.

Today's Smart Parks: I just returned from a walking tour of the Dorothea Dix property in downtown Raleigh, North Carolina. This 308-acre land development project is one of the most interesting and dynamic park initiatives in the United Sates. It could become Raleigh's Central Park. The master plan bills it as an economic engine and a world-class example of civic engagement (both key properties of a smart park). On my ninety minute, three-mile ramble around what is known as Dix Hill, I discovered a large swath of land that would offer breathing space to the nearby downtown area that looks (basically) squeezed into a few square blocks. I couldn't help thinking about ways the City would restore, manage, maintain and promote the use of this large urban park. How would the City of Raleigh differentiate itself from other cities with successful park initiatives? After all, the purchase and retrofitting of this vast piece of land is a large and expensive undertaking for the City of Raleigh.

I recently spoke with the City of San Francisco about the role of geospatial technology in park evaluations. Monitoring the condition of parks and park infrastructure is often a broad-brush strategic evaluation that uses very specific tools. These evaluations are very different beasts than full park asset inventories.

Much to my delight, I also discovered that the university city of Berkeley, California, is about to also embark on a smart park inventory that embraces all components of GIS technology. This is a fundamental prerequisite for a smart park. The City of Irvine and Orange County, California are addressing the need for geospatial technology in their new Great Park, an innovative public space of 1,300 acres of amenities and park space. More than 450 acres of park space has been developed, with more than 230 additional acres in progress. There is much afoot in the world of parks and recreation. There is no better time for this industry to examine smart practices.

We are now entering the era of smart parks. This term developed in parallel with the smart cities initiatives taking hold around the world. But what is a smart park? Based on conventional wisdom, a smart park is a **park that uses advanced technology to reach a series of common goals**. Deploying effective technology in parks provides a clear benefit to the public while additionally improving operations, management and programming. Also according to National Recreation and Parks Association (NRPA) "Creating technology-enabled smart parks does not require a monumental shift in management practices or departmental organization, nor does their creation require excessive capital expenditure. Quite the opposite. Start small and build on success with incremental changes." Making Smart Parks, May 1, 2016, Feature by Edward Krafcik (Krafcik, 2016).

Smart parks use technology to improve park performance, advance social equity, increase park access, enhance community health and well-being, improve safety and resilience to climate change, improve water and energy usage and ease operation and maintenance. Smart parks should follow principles of a smart organizations: allowing **personalized recognition between people and systems** (for example, bike and scooter sharing); **providing location-based information** (on-demand buses); **deploying geo-sensors** that observe, understand and anticipate events (air quality, noise, climate,

---

Smart Park Goals and Objectives

- Social Equity
- Community Health and Fitness
- Safety
- Resilience
- Sustainability
- Water Efficiency
- Storm Water Management
- Energy Efficiency
- Effective Operations and Maintenance
- Internet Connectivity / Wi-Fi
- Community Building
- Improved Social Trust
- Effective Maintenance
- Improved Air Pollution

FIGURE 7.19
Smart Park Goals and Objectives.

pedestrian use); **linking people** to services, resources, amenities and each other (weather feeds, traffic counts, Tweets and other social media); and the **interoperability** that enables different systems, information sources and data types to work together in order to create new decision-support information.

We must understand the smart park vision and goals and objectives of an organization then subsequently meet those needs with technology. Figure 7.19 details common smart park goals and objectives.

The Luskin Center of Innovation created a smart parks toolkit detailing a great deal of new technology that is being used in parks around the world. Figure 7.20 has been created from a number of sources including the Luskin Center toolkit.

Parks are essential to our villages, towns, cities and counties. Their beneficial relationship with community health, environmental, community and economy is well documented. Tools to measure the positive impact of parks and recreation are critical to our future well-being. If it can't be measured, it doesn't exist!

Geospatial Technology and GIS

So where does geospatial technology and GIS fit in to all of this smart park technology? Parklands, recreation, open space, critical habitat and natural

Smart Park Technologies			
The Luskin Center of Innovation created a Smart Parks Toolkit that details a lot of new technology that can be, or is being, used in parks around the world. The following table has been created from a number of sources, including the Luskin Center Toolkit.			
Landscape Technologies	**Irrigation Technologies**	**Storm Water Technologies**	**Hardscapes and Impervious Surface Technologies**
• Automated lawn mowers • Near-infrared photography • Green roofs • Green walls • Air pruning plant containers • Vibrating pollinators	• Smart water controllers • Low-pressure and rotating sprinklers • Subsurface drip irrigation • Smart water metering • Greywater recycling	• Engineered ssoils • Underground storage basins • Drones • Real-time control and continuous monitoring and adaptive control • Rainwater harvesting	• Paved walking paths • Parking lots • Seating areas • Cross-laminated timber • Pervious Parking • Piezoelectric energy harvesting tiles • Self-healing concrete • Photocatalytic titanium dioxide coating • Transparent concrete • Daylight fluorescent aggregate carbon upcycled concreate
Active Space Technologies	**Furniture and Amenities Technologies**	**Lighting Technologies**	**Digi-scapes Technologies**
• Interactive play structures • High performance track surfaces • Pool ozonation • Energy-generating exercise equipment • Outdoor DJ Booths • Hard surface testing equipment	• Smart benches • Solar shade structures • Solar powered trash compactors • Restroom occupancy sensors • Automated bicycle and pedestrian counters	• Motion-activated sensors • LED's and fiber optics • Off-grid lighting fixtures • Digital additions to LED fixtures • Lighting shields	• Wi-Fi • Application software • Sensor networks • Internet of Things

FIGURE 7.20
Smart Park Technologies.

resources within our communities are vital to our future. We need to understand the assets of our parks, how we interact with our natural landscape and how our communities recreate. We need to know everything about the natural, green infrastructure within our communities and our corresponding patterns of human interaction.

Understanding Park Resources

How many types of parks do we have in our community? How much open space do we have in relation to our population? How much environmentally sensitive area exists? What is the type and condition of all our green infrastructure, from swings and slides to wetlands and woodland? Are we growing our parkland, or slowly losing it to urban growth? GIS offers mobile and office data-based analytical solutions to inventory, quantify, analyze and maintain a real-time database of our park lands and park infrastructure. Today, it's all about data analytics! Real and accurate data will drive policy.

Understanding Recreation Patterns and Trends: People and Community

Understanding demographic patterns and recreation participation trends within a community will help decision-makers plan and build for the future. Knowing the what, where, who, and how of a local situation will play a significant role in managing recreation services and activities. Where do all our active and registered recreation users live? How many people come from outside our city limits to use the recreation facilities and why? Are there any social equity gaps in our system? How long does it take for any resident to walk to a park? Which neighborhoods are more active than others? Do we have the right balance of parks, open space, people and recreation activities? How well are we serving our community? What is the relationship between our citizens and our parks and recreation solutions?

Understanding Recreation Patterns and Trends: Social Equity and Social Economics of Parks and Infrastructure

Lower-income communities and communities of color tend to have less access to quality parks and recreation facilities. Evidence shows that racial disparities exist regarding accessibility to recreation facilities! Parks and recreation facilities provide the most important opportunities for physical activity and play. GIS introduces tools that illustrate these social equity issues and support effective design of solutions that improve accessibility to parks and recreation facilities. What is the value of a tree to our community? What is the value of our parks and open space? How can we investigate and examine parkland user activity, our demographics and social fabric?

Strategic and Sustainable Management

GIS offers a management solution that centers on improving efficiency, reducing costs, increasing productivity, saving time, increasing our understanding of our natural landscape and promoting collaboration and sustainability. GIS allows us to understand the inter-relationship of green infrastructure and our management components, including:

- **Planning**: What's the ideal location for a new park? What site serves our community best? How can we effectively connect linear greenways? How difficult would a new walking trail be? What is the economic impact of our parks? How can we use social media to understand the future of our parks and recreation? How will socio-economic trends influence our decisions? Where is the best place to put recycle containers? How do we create more equitable communities? What information do we need to plan for the future? How can parks and recreation departments address the needs of the community?

- **Operations**: What is the condition of all of our green infrastructure? How many park benches do we own? How much open space needs mowing? How long should it take for field crews to mulch an area? How much infrastructure needs repairing next year? How much impervious surface do we have in our parks? How much should we budget for improvements? How much should we request for capital improvements? How much work can we complete in a week? Where should we improve lighting in the park?

- **Policy**: People who live near parks are more likely to be active. Youths in neighborhoods with recreation facilities are more likely to be active than their counterparts. Every dollar spent on recreation trails correlates to a significant savings in direct medical costs. Homes near parks sell for more money. Sharing inter-governmental recreation resources benefits the entire community. Parks and recreation programs help reduce childhood obesity! The economic benefits of open space are self-evident. GIS empirically highlights facts that will shape future policy decisions.

A Bespoke Solution for Parks and Recreation

A bespoke solution is a tailored group of applications that support all geospatial aspects of parks and recreation.

1. **Map Viewer**

   This is an intuitive solution for viewing, querying and analyzing parks and recreation data. This tool offers new and innovative decision support and solutions that will drive the Parks and Recreation Master Planning process. Viewing, interpreting, analyzing and monitoring park infrastructure and recreation users has never been easier (Figure 7.21).

   - A real-time view of your parks system and recreation users
   - Demographic information and analysis

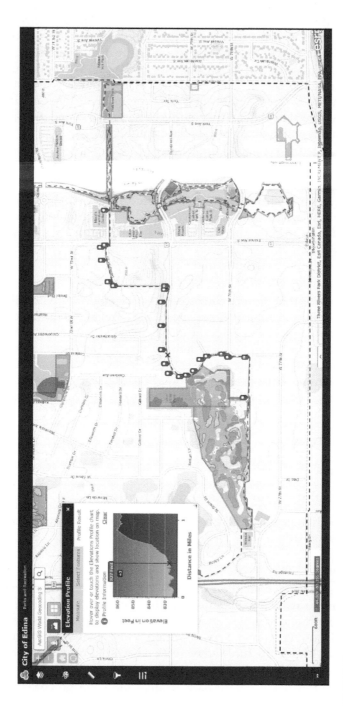

FIGURE 7.21
Map Viewer.

- Geo-analytics to identify gaps, opportunities, strengths and patterns
- Develop software widgets and tools for planning and decision support

2. **Field Data Collector: Survey 123**

An easy-to-use map-centric solution to inventory, update and maintain your park system infrastructure and perform field surveys. It is a mobile solution that works on any of your hardware – laptops, tablets, smart phones. This technology makes the task of collecting park infrastructure and recreation user data in the field and turning it into meaningful information straightforward and easy (Figure 7.22).

- Easy to use Graphic User Interface (GUI)
- Works on tablets and smartphones
- Touchscreen intuitive technology
- Simple search and query functionality
- Add photographs, video and field notes

3. **Operations Dashboard**

A unique and interactive operations dashboard that dynamically monitors park assets, recreation programs and events. It is a new solution for monitoring the status and performance of parks and recreation departments. The operations dashboard easily tracks field activities and keeps an up-to-date record of characteristics of park-land, open space, wooded land, wetlands, parking facilities, the condition of playground and park infrastructure equipment, and any user-defined metrics (Figure 7.23).

- Real-time park system and recreation user status
- Dashboard decision support
- Operational performance measures
- Existing condition assessment
- Measuring policy and standards

FIGURE 7.22
Field Data Collector.

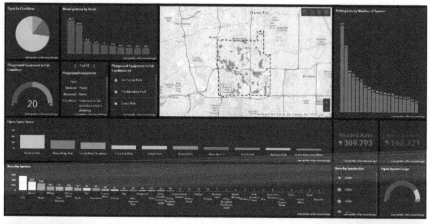

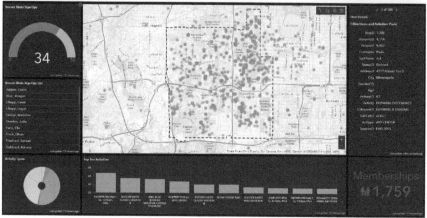

FIGURE 7.23
Operational Dashboard.

4. **Resident and Citizen Awareness and Engagement**

   This solution improves community awareness and engagement through geospatial story maps, walking tours an online citizen park and recreation query tool. Public facing applications offer an exciting and visually pleasing way of engaging citizens (Figure 7.24).

   - Increase community awareness
   - Promote parks and recreation opportunities
   - Showcase parks
   - Improve public access

5. **Dynamic Integration – Interoperability of Any Management Database**

   This solution offers a real and present opportunity to integrate data from many existing software solutions. This data can include

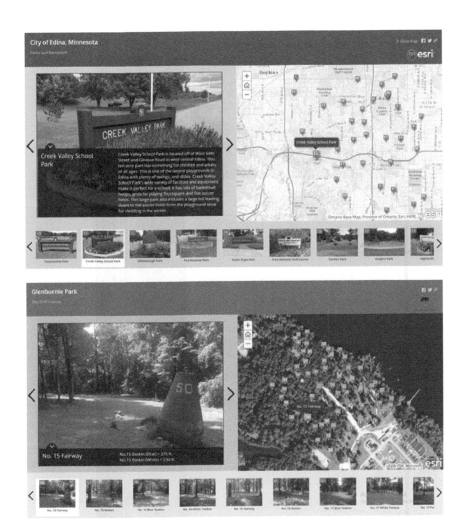

FIGURE 7.24
Residential Awareness – Story Maps.

real-time recreation user information, work order and work request data, park user survey data, tree inventory systems, and revenue data (Figure 7.25).

- Integrate real-time recreation user information
- Analyze recreation facility usage and performance
- Increase park and recreation participation and revenue
- Locate all your park and recreation users
- Determine area to market recreation events, activities and facilities

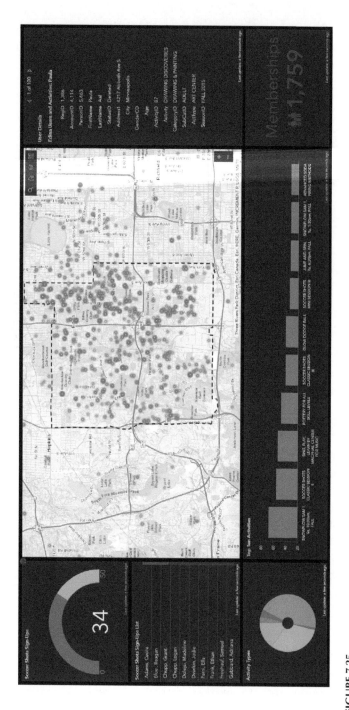

FIGURE 7.25
Dynamic Database Integration.

- Make informed decision about recreation user habits and patterns
- Use census data to understand your community demographics

Business Realization Planning for Smart Park Initiative

Hoffman Estates Park District invested time and money into a smart GIS park initiative. Mr. Gary Buczkowski, Director Planning and Development at Hoffman Estates Park District developed a paper entitled "Return on Investment Analysis Secures Funds for GreenCityGIS Initiative" that managed to secure the funding necessary for deploying a bespoke solution. Buczkowski explained,

> Directing limited resources toward the best end use return on investment has become the number one priority of successfully run agencies. And while the Hoffman Estates Park District has had success in the past, there is no guarantee that in the future this will continue especially given the fact that the agency has little to no new growth potential. Managing what we have will be the priority going forward.
>
> **– Buczkowski**

GIS system along with other data base software systems that are specially tailored for the parks and recreation industry may just be the key to operational success in the future. How can GIS benefit the Hoffman Estates Park District in future decision making?

- Cost savings from greater efficiency
- Better decision making GIS is the "go-to-technology" for making better decisions about location
- Better communicative tools with relative reliable data
- Record keeping

8

Conclusion

I am more and more convinced that our happiness or unhappiness depends more on the way we meet the events of life than on the nature of those events themselves.

– Alexander von Humboldt

Geo-Spatial Technology, Performance, Value, Culture and Software Applications

Geo-Spatial Technology: Geo-Smart Government
Connected Technology Ecosystem – A System of Systems

In thirty years, the world's population will be nearly 10 billion people, out of which 6.7 billion will live in cities. This rapid population growth will increase the demand for employment, housing, energy, clean water, food, transportation, health, education, social services, social equity and safety.

The world of municipal government is not only becoming more urban; it's becoming far more complex. Our towns, cities and counties will be continuously asked to improve efficiency, increase productivity and make informed technologically based decisions. They will be asked to find new ways to protect the community, improve citizen access, comply with state and federal mandates, respond and interact quicker with citizens, process and interpret social media data and openly, transparently and effectively manage our resources.

Our cities must become ever more resilient. To face the coming challenges, cities must be able to survive, adapt and grow in the face of economic, social and physical challenges. Local governments would be wise to take pointers from sustainable and smart communities like Copenhagen, San Francisco, Vancouver and Stockholm where citizens feel safe, healthy and protected because of their city's commitment to greenness, carbon neutrality, energy efficiency, conservation efforts, clean energy, sustainable food sources and quality of life.

To meet the demands of – what seems to be – uncontrollable urban population growth, local governments will need to think carefully about technology and management styles that relate to smart cities, performance, value and culture? **Geospatial science is not only helping government meet these demands; it's redefining how they do it.**

If we look back at the history of geospatial technology, we can see distinct patterns of evolution. If you look closely, you can see the GIS professional as an agent of change. Pioneers, manager and coordinators have all affected the GIS industry. Soon, the GIOs of the future will manage and promote a very different geospatial world.

GIS has matured in a very predictable way. Each stage of GIS shares similar characteristics with the preceding stage; only these characteristics are amplified by a nudge toward a new direction. It may seem that advances are made overnight, but they never are! It always takes time. Sometimes up to thirty or forty years.

If we compartmentalize the last sixty years into four distinct evolutionary phases of geospatial technology, we have:

- 1960 to 1980: Proof of Concept: **GIS Pioneers**
- 1980 to 2000: Desktop, Analysis, and Projects: **GIS Managers**
- 2000 to 2020: Enterprise, Strategic, and Scalable: **GIS Coordinators**
- 2020 to 2040: Smart, Resilient and Sustainable: **Geographic Information Officers (GIO)**

In 1970, Sir Fred Hoyle, an English astronomer and cosmologist, who studied science at Emmanuel College, Cambridge, stated,

> "You will have noticed how quite suddenly everybody has become seriously concerned to protect the natural environment. Where did this idea come from? You could say from biologists, conservationists and ecologists. But they have been saying the same thing as they are saying now for many years. Previously they never got on base. Something new has happened to create a world-wide awareness of our planet as a unique and precious place. It seems to me more than a coincidence that this awareness should have happened at exactly the moment Man took his first step into space."

We can see our future not as a series of dramatic paradigm shifts but as a slow and purposeful technological evolution. If we listened carefully to the pioneers of GIS technology – or even the GIS managers of the 1990s – we would have them discussing data sharing as a key future goal. We would have been able to anticipate – not predict – that our contemporary focus would center on **open and transparent government, the instant feedback from residents through social media, online citizen engagement, neighborhood networks, citizen crime solutions, government scorecards** and **citywide livable index dashboards.**

There is no psychic energy, revelation, foreshadowing or prophecy in play here. Recall what I said in the first chapter: **the future is always the present. We react, we do not predict.** The same goes for technological shifts. Over the past thirty years we have improved upon existing ideas and prior theories. We have not re-evaluated prior facts nor have we reconstructed prior theories. In 1988, Mr. Jack Dangermond, President of Esri said, "GIS technology offers great promise for improving the future of billions of people."

We will see Mr. Dangermond's words bear fruit well into the future. A premium ArcGIS HUB will be the IT managed, cloud-based oracle of data gathering, analysis, interpretation, dissemination and story boarding. The Internet of Things (IoT), sophisticated data analytics and Artificial Intelligence (AI) will walk front and center in the march forward.

An Established Footing: If we are to successfully understand and potentially accelerate the growth of geospatial technology in local government, we need a foundation. We need an established footing. In the interest of "simplifying the complicated," we must grasp the components of GIS, understand the evolutionary stages of GIS in government and comprehend the concepts of smart cities and high performance organizations. A firm understanding what GIS and geospatial science can do in terms of applications is of foundational importance. The purpose of this book is straightforward. It is a conversation about seven important "working gears" of GIS. They include:

1. Six Pillars of GIS Sustainability
2. Six Logical Stages of GIS Maturity
3. The Smart City of the Future
4. High-Performance Organizations (HPO)
5. Business Realization Planning (BRP)
6. Sustainability and Resilience
7. Applications in Local Government: A Smart Geospatial Ecosystem

The evolution, maturity and competence levels of GIS in local government can be determined and documented using two specific methods of measurement. They include the Six Pillars of GIS Sustainability and the Six Logical Stages of GIS Maturity. The **Six Pillars of GIS Sustainability** attempts to describe and discuss the future of the geospatial industry and geo-smart government by deconstructing the question "What is GIS?" into the six pillars of GIS sustainability.

Pillar 1: Smart Geospatial **Governance** Components

Pillar 2: Smart **Digital Data and Databases** Components

Pillar 3: Smart **Procedures, Workflow and Interoperability** Components

Pillar 4: Smart **GIS Software** Components

Pillar 5: Smart **Training, Education and Knowledge Transfer** Components

Pillar 6: Smart GIS **IT Infrastructure** Components

The **Six Logical Stages of GIS Maturity** are fundamentally tied into the strategic, technical, tactical, logistical and political challenges that exist in local governments at every stage of GIS maturity. The six stages of maturity include:

Stage One: Adoption

Stage Two: Enhanced

Stage Three: Operational Efficiency

Stage Four: Strategic and Operational Efficiency

Stage Five: Enterprise, Strategic and Operational Excellent

Stage Six: Smart and Sustainable – Smart City

The smart city of the future is a designation given to a local government organization that creates a "connected city." A "connected city" incorporates, integrates and embeds information technology into the decision-support apparatus of local government. It is a city that works toward an interoperable system of systems and embraces both big data analytics and predictive analysis. The smart city explores new ways of consuming and exploring disparate databases. It grasps the importance of comprehending new relationships between data. A smart community is a municipality that focuses on the real-time nature of data in service of increasing operational efficiency, sharing information across an organization and improving the quality of government services.

Performance: High-Performance Organizations
Organic and Continuous Management and Improvement

High-performance organizations (HPO) are predicated upon organic and continuous improvement. Local government must examine the ways a HPO management strategy can be used for geospatial technology in local government. The HPO is a conceptual, scientifically validated structure that managers use to decide what to focus on in order to improve organizational performance and sustainability. An HPO framework, translated by managers for the specific needs of their organization, provides us with a solid recipe for success. The future is HPO. A successful HPO will plan, design and implement state-of-the-art geospatial applications and solutions.

Value: Business Realization Planning
Measuring Key Performance and Business Functions

The Business Realization Plan (BRP) encompasses the business case, appropriate measures, benefit drivers, processes and ongoing monitoring of the benefits provided by geospatial technology to the organizations. A BRP provides us with a comprehensive evaluation of geospatial technology in local government. It serves a way to measure how projects and programs true value to organizations and provides a set of questions and practices that local government management professionals can use to help guide the identification, analysis, delivery, and sustainment of benefits that align to the organization's strategic goals and objectives. The Project Management Institute (PMI) Thought Leadership (2016) support Business Realization Planning as a core strategic initiative in local government.

Culture: Sustainability and Resilience
Environmental, Economic and Social

Remember what Michio Kaku said in his book *The Physics of the Future – How Science Will Shape Human Destiny and Our Daily Lives by the Year 2100*. Kaku stated that "we will rank civilization by its energy efficiency or entropy" (gradual decline into disorder). This process has already begun. Its warms the heart to see the American Council for Energy Efficient Economy (ACEEE) presenting energy efficiency scorecards that rank states and cities on a number of criteria including, local government operations, community wide initiatives, building policies, and energy and water utilities and transportation.

Sustainability in Local Government: The phrase "sustainable communities" is used synonymously with the terms "green cities," "livable cities" or "smart cities." Sustainable communities are places where local government meets the needs of everyone in the community regarding safety, health and the quality of life. Tomorrow's sustainability strategies will focus on more than just wind energy, solar power, sustainable construction, green space and sustainable forestry. GIS applications will be at the forefront of measuring sustainability.

Resilience in Local Government: A resilient community is one that has developed the capacities to absorb future events, shocks, and trauma to its social, economic, technical and infrastructure systems. The resilient city is able to maintain its functions, structures and systems in the face of such shocks. The City of Berkeley in California offers a good example of a city that developed and adopted a resilience plan. Unfortunately, I'm not sure that developing a resiliency plan at the local government level is not yet common practice. I am sure that GIS will play the leading role in future resilience planning.

Software Applications: Applications in Local Government:
A Smart Geospatial Ecosystem

The smart GIS ecosystem of software will only expand its significance within in local government. In whatever form it takes, GIS software will continue to change the way every local government department operates. Today there are three specific application areas:

- Desktop Solutions
- Web Solutions
- Mobile Solutions

Each system of "geospatial delivery" offers abundant opportunities to change the way local government thinks and operates. The natural evolution of GIS is happening in front of our eyes. We are moving toward **a ubiquitous and extensive ecosystem** that includes everything from sophisticated desktop functionality to web-based architecture to mobile solutions.

Recall that currently there are two main types of software initiatives in local government: Enterprise Software Solutions that are used by and for the entire organization and Departmental Software Solutions that are used by specific departments. In this book, we looked at the sixteen applications commonly used in local government.

- My Government Services
- Parcel Notification
- Arc GIS HUB –Open Data Sharing and Transparent Government
- Crowdsourcing Applications
- Crime Analysis
- Emergency Operations Center (EOC) Browser
- Emergency Operations Center (EOC) Dashboards
- Story Map – Cold Case
- Story Maps – Flooding
- Utilities Web Browser
- Utilities Dashboard
- Fiber Management Solution (FMS)
- Fiber Analysis Web Browser
- Work Force Solution
- Economic Web Application
- Planning and Zoning – 3D Urban Solutions

Figure 8.1 illustrates some important GIS applications used in local government today.

Today's robust mapping and analytics platforms will springboard our communities into the world of high-tech government. GIS technology is the engine that allows cities to become smart, sustainable, resilient and high performing. As we anticipated forty years ago, we can collect field data at lightning speed, perform spatial analysis within minutes and

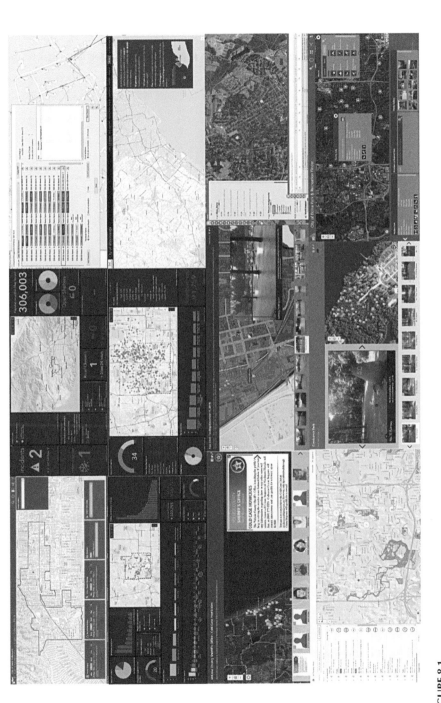

FIGURE 8.1
Collage of software application in local government.

visualize then present an understandable story about current events at a moment's notice.

Our challenge is to understand the growing geospatial ecosystem of software and solutions. Indeed, we must learn "An Altogether Different Language."

References

Bateson, Gregory. *Mind and Nature*. Collins, Glasgow, 1980.

"Benefits Realization Management Framework." Project Management Institute, 2016.

Benfield, Kaid. "What Does a 'Sustainable Community' Actually Look Like?" *The Atlantic*, Atlantic Media Company, 14 Mar. 2011, www.theatlantic.com/national/archive/2011/03/what-does-a-sustainable-community-actually-look-like/72376/.

"Berkeley." 100 Resilient Cities, www.100resilientcities.org/cities/berkeley/.

Buczkowski, Gary. "Bid Information." Hoffman Estates Park District, 2019, www.heparks.org/general-information/bid-information/.

"Business Realization Planning Framework." PMI Thought Leadership Series, www.pmi.org/-/media/pmi/documents/public/pdf/learning/thought-leadership/benefits-realization-management-framework.pdf.

"Certification." Parksmart, parksmart.gbci.org/certification.

"The Characteristics of a High Performance Organisation." André De Waal, www.andredewaal.eu/publication/the-characteristics-of-a-high-performance-organisation-2/.

Chen, James. "Return on Investment (ROI)." Investopedia, Investopedia, 12 Mar. 2019, www.investopedia.com/terms/r/returnoninvestment.asp.

City of Mississauga. "Smart City Master Plan: A Smart City for Everybody." June 2019, https://www7.mississauga.ca/websites/smartcity/SMRTCTY_Master_Plan_Final.pdf.

"City of St. Louis: The Smart City Challenge Application Connecting People and Opportunity Vision Narrative." U.S. Department of Transportation, 4 Feb. 2016.

"City Resilience Program." World Bank, www.worldbank.org/en/topic/disasterriskmanagement/brief/city-resilience-program.

Cloke, Ken and Joan Goldsmith. *The End of Management and the Rise of Organizational Democracy*. Jossey-Bass, San Francisco, 2002.

Collins, James C. *Good to Great*. Random House Business, London, 2001.

"Commonwealth Center for High Performance Organizations." Commonwealth Center for High Performance Organizations, www.highperformanceorg.com/.

Dangermond, J. and L. K. Smith. Geographic Information Systems and the Revolution in Cartography: The Nature of the Role Played by a Commercial Organization. *The American Cartographer*, vol. 15, 1988, p. 309.

Dempsey, Caitlin. "History of GIS." *GIS Lounge*, 12 Jan. 2018, www.gislounge.com/history-of-gis/.

"Done and Dusted." Not One-Off Britishisms, Wordpress.com, 6 Oct. 2015, https://notoneoffbritishisms.com/2015/10/05/done-and-dusted/.

"New ICMA Home Page." Icma.org, icma.org/.

Eschwass. "The State Energy Efficiency Scorecard." *ACEEE*, 4 Oct. 2018, aceee.org/state-policy/scorecard.

"Esri." GIS Mapping Software, Spatial Data Analytics & Location Intelligence, www.esri.com/en-us/home.

Florence, Curtis S., et al. "The Economic Burden of Prescription Opioid Overdose, Abuse, and Dependence in the United States, 2013." *Medical Care*, vol. 54, no. 10, 2016, pp. 901–906, doi:10.1097/mlr.0000000000000625.

Fotheringham, Stewart and Peter Rogerson. *Spatial Analysis and GIS*. Taylor & Francis, Philadelphia, PA, 2014.

Fountaine, Tim, et al. "The AI-Powered Organization: The Main Challenge Isn't Technology. It's Culture." *The Harvard Business Review*, 2019.

"Free Driving Directions, Traffic Reports & GPS Navigation App by Waze." Free Driving Directions, Traffic Reports & GPS Navigation App by Waze, www.waze.com/.

"Getting to Green: Paying for Green Infrastructure Financing Options and Resources for Local Decision-Makers." USEPA, 2014, https://nepis.epa.gov/Exe/ZyPDF.cgi?Dockey=P100LPA6.txt.

"Greenlight Community Broadband." Greenlight Community Broadband | Home Greenlight, www.greenlightnc.com/.

"The High Performance Organization (HPO) Framework." HPO Center, www.hpo-center.com/hpo-framework/.

Holbeche, Linda. *High Performance Organization*. Routledge, London, 2015.

Holdstock, David A. "GIS: Yesterday, Today, and Tomorrow." Elevations Geospatial Summit. Elevations Geospatial Summit, June 2018, Wyoming.

Holdstock, David A. *Strategic GIS Planning and Management in Local Government*. CRC Press, Boca Raton, FL, 2017.

"Home." ShotSpotter, www.shotspotter.com/.

"Home Page." 100 Resilient Cities, www.100resilientcities.org/.

Hoyle, Fred. "After Dinner Speech." NASA First Lunar Sciences Conference. 6 Jan. 1970, Houston.

Huxhold, William E. *An Introduction to Urban Geographic Information Systems*. Oxford University Press, 1991.

Infrastructure Canada. "Smart Cities Challenge." Infrastructure Canada, 18 Apr. 2019, www.infrastructure.gc.ca/cities-villes/index-eng.html.

"IPv6 Basics, News, Guides & Tutorials." Internet Society, www.internetsociety.org/deploy360/ipv6/.

Kaku, Michio. *Physics of the Future: How Science Will Shape Human Destiny and Our Daily Lives by the Year 2100*. Anchor Books, New York, 2012.

Kenton, Will. "How Cost-Benefit Analysis Process Is Performed." Investopedia, Investopedia, 23 June 2019, www.investopedia.com/terms/c/cost-benefitanalysis.asp.

Krafcik, Edward. "Making Smart Parks." The National Recreation and Parks Association, 1 May 2016, www.nrpa.org/parks-recreation-magazine/2016/may/making-smart-parks/.

Kuhn, Thomas S. *The Structure of Scientific Revolutions*. The University of Chicago Press, 2015.

Kushner, Lawrence. Invisible Lines of Connection. Jewish Book Council, New York, 1998.

LaDuca, Ann and John Kosco. "Getting to Green: Paying for Green Infrastructure Financing Options and Resources for Local Decision-Makers." 2014, Getting to Green: Paying for Green Infrastructure Financing Options and Resources for Local Decision-Makers.

Levin, G. Benefits – A Necessity to Deliver Business Value and a Culture Change but How Do We Achieve Them? Paper presented at PMI® Global Congress, 2015, North America, Orlando, FL, Newtown Square, PA: Project Management Institute.

Loukaitou-Sideris, Anastasia. "Smart Parks: A Toolkit." UCLA Luskin School of Public Affairs, Luskin Center for Innovation, innovation.luskin.ucla.edu/sites/default/files/ParksWeb020218.pdf.

Maddox, Teena. "Smart Cities: 6 Essential Technologies." TechRepublic, Tech Republic, 25 Oct. 2017, www.techrepublic.com/article/smart-cities-6-essential-technologies/.

Maxwell, John C. *Good Leaders Ask Great Questions: Your Foundation for Successful Leadership.* Center Street, New York, 2016.

McArthur, G., et al. "Oxford English." Oxford English, Oxford University Press, 1993.

Newton, Isaac. "Letter to Robert Hooke." 5 Feb. 1676.

"1000 GIS Applications & Uses - How GIS Is Changing the World." GIS Geography, 4 Mar. 2019, gisgeography.com/gis-applications-uses/.

Parker, Dorothy, Great Hollywood Wit compiled and edited by Gene Shalit p. 96.

Pickering, John W. *Building High-Performance Organizations for the Twenty-First Century.* Commonwealth Center for High Performance Organizations Inc., 1993.

"PMIS Consulting Limited." PMIS Consulting Limited, www.pmis-consulting.com/.

Porter, Anne and Anne Porter. *Living Things: Collected Poems.* Zoland Books, 2006. "An Altogether Different Language."

"Protection at Every Corner." Ring, ring.com/.

Reed, Carl. "Geospatial Paradigm Shift or Not? Featured Article By Carl Reed, Chief Technology Officer and Executive Director Specification Program, OGC." *GISCafe*, 6 Sept. 2005, www10.giscafe.com/nbc/articles/view:article.php?articleid=203579.

"Ring Video Doorbell." Wikipedia, Wikimedia Foundation, 13 June 2017, en.wikipedia.org/wiki/Ring_Video_Doorbell.

Roosevelt, Theodore. "Citizenship in a Republic." 23 Apr. 1910, Paris, Sorbonne.

Saunders, Doug. "In 100 Years We Will Be an Entirely Urban Species." *The Spectator*, 7 Aug. 2010, www.spectator.co.uk/2010/08/in-100-years-we-will-be-an-entirely-urban-species/.

Scott, Ridley, Drew Goddard, Simon Kinberg, Michael Schaefer, Aditya Sood, Mark Huffam, Matt Damon, Jessica Chastain, Kristen Wiig, Jeff Daniels, Michael Peña, Kate Mara, Sean Bean, Sebastian Stan, Aksel Hennie, Mackenzie Davis, Benedict Wong, Donald Glover, Shu Chen, Eddy Ko, Chiwetel Ejiofor, Harry Gregson-Williams, Pietro Scalia, Dariusz A. Wolski, and Andy Weir. The Martian, 2015.

"The Secret of High Performance Management Organizations." André De Waal, www.andredewaal.eu/publication/the-secret-of-high-performance-management-organizations/.

Siemens' Green Cities Index, etms.espon.eu/index.php/this-big-city/qr/534-siemens-green-cities-index.

Taylor, Frederick W. *The Principles of Scientific Management.* Harper & Brothers, New York, 1911.

"Uber (Company)." Wikipedia, Wikimedia Foundation, 25 Feb. 2019, en.wikipedia.org/wiki/Uber_(company).

Varghee, Romy. "America's Cities Are Running on Software from the '80s." Bloomberg. com, Bloomberg, 8 Feb. 2019, www.bloomberg.com/news/articles/2019-02-28/ america-s-cities-are-running-on-software-from-the-80s.

Waal, André de. *What Makes a High Performance Organization: Five Validated Factors of Competitive Advantage That Apply Worldwide*. Warden Press, Amsterdam, 2019.

"Waze." Wikipedia, Wikimedia Foundation, 8 July 2019, en.wikipedia.org/wiki/Waze.

"What Is XaaS (Anything as a Service)? - Definition from WhatIs.com." Search CloudComputing, searchcloudcomputing.techtarget.com/definition/XaaS-anything-as-a-service.

Wikipedia Contributors. "Benefits Realisation Management." Wikipedia, The Free Encyclopedia. Wikipedia, The Free Encyclopedia, 9 Jun. 2019. Web. 12 Aug. 2019.

Wikipedia Contributors. "Risk–Benefit Ratio." Wikipedia, The Free Encyclopedia. Wikipedia, The Free Encyclopedia, 19 Apr. 2018. Web. 12 Aug. 2019.

Index